I0815117

BARRY B. KAPLAN

HOROLOGY

An Illustrated Primer on the History, Philosophy, and Science of Time, with an Overview of the Wristwatch and the Watch Industry

SCHIFFER PUBLISHING®

4880 Lower Valley Road • Atglen, PA 19310

Copyright © 2022 by Barry B. Kaplan
Unless otherwise indicated, all materials on these pages are copyrighted. All rights reserved. No part of this publication, including text or image, may be reproduced for any purpose other than personal use. Therefore, except for brief quotations in a book review, reproduction, modification, storage in a retrieval system, or retransmission, in any form or by any means, electronic, mechanical, or otherwise, for reasons other than personal use, are strictly prohibited without prior written permission.

Library of Congress Control Number: 2021948661

All rights reserved. No part of this work may be reproduced or used in any form or by any means—graphic, electronic, or mechanical, including photocopying or information storage and retrieval systems—without written permission from the publisher.

The scanning, uploading, and distribution of this book or any part thereof via the Internet or any other means without the permission of the publisher is illegal and punishable by law. Please purchase only authorized editions and do not participate in or encourage the electronic piracy of copyrighted materials.

"Schiffer," "Schiffer Publishing, Ltd.," and the pen and inkwell logo are registered trademarks of Schiffer Publishing, Ltd.

Edited by Ann Charles
Designed by Justin Watkinson
Type set in Proxima Nova/Minion Pro

ISBN: 978-0-7643-6392-4
Printed in China
6 5 4 3

Published by Schiffer Publishing, Ltd.
4880 Lower Valley Road
Atglen, PA 19310
Phone: (610) 593-1777; Fax: (610) 593-2002
Email: Info@schifferbooks.com
Web: www.schifferbooks.com

FSC
www.fsc.org
MIX
Paper | Supporting responsible forestry
FSC® C167893

Schiffer Publishing's titles are available at special discounts for bulk purchases for sales promotions or premiums. Special editions, including personalized covers, corporate imprints, and excerpts, can be created in large quantities for special needs. For more information, contact the publisher.

Legal disclaimer
All company, brand, collection, product, part, and material names are trademarks (™) or registered trademarks (®) of their respective holders. Use of these names does not imply any affiliation with or endorsement by them. The advice and strategies found within may not be suitable for every situation. The content provided is for educational purposes only and does not take the place of professional advice. Every effort has been made to ensure accuracy at publishing time. However, this is not an exhaustive treatment of the subject. No liability is assumed for losses or damages due to the information provided. You are responsible for your own choices, actions, and results.

בס״ד

To Lance, the finest of brothers, and the finest of people—I miss you.

To Grandpa Barney, a humble Swiss watchmaker. Your passion lives on.

And, to Mr. Ted Kaplan, my beloved father, a mensch of blessed memory, I miss you. I hope this book makes you proud.

Contents

WHAT IS THIS BOOK?

"If one cannot enjoy reading a book over and over again, there is no use in reading it at all."

—Oscar Wilde

Combining *hora*[1] (hour or time) with *logy* (the study of), the noun *horology* describes the study and measurement of time and the art of making clocks and watches.

Many books discuss wristwatches, their fabrication, and the history of timekeeping. Those that focus specifically on the discipline of watchmaking usually tackle setting up a watchmaking workshop or fashioning pinions, wheels, or bridges. Others focus on the formation and progress of specific watch brands, with accompanying images illustrating artifacts from the brands' archives and elsewhere. Still others examine time exclusively from a philosophical, sociological, or scientific perspective. Then there are those books that are horological eye candy—wristwatch photo books or annuals akin to glorified catalogs (not that there is anything wrong with that), with or without pricing or technical information.

This book incorporates each of the aforementioned themes and expands further, to all areas of watchmaking, including design, decorative techniques, starting a watch collection, background to the watch industry, the history of the calendar, and much more.

The shared passion for all things automotive is common among watch lovers. The intricacy of a multicylinder engine (with or without forced induction) working in tandem with a well-honed chassis, intuitive body control, and gorgeous bodywork has a one-for-one analog in the watch world. It is no wonder that over the decades, various watch brands have spent exorbitant marketing dollars on professional motorsports. Throughout the book I reference automotive themes in addition to other topics and disciplines of related interest. This is not just a book about watches. It is, one hopes, much more than that.

A compendium of sorts, this volume is an overview of the captivating world of time, watchmaking, and kindred disciplines. The diligent novice should, I hope, gain a useful grounding in the fundamentals. Experienced horologists will already know a great deal about the material, but they might gain insight into ancillary topics and deeper understanding elsewhere. It is part coffee-table book, part technical reference, part socioeconomic analysis, and part philosophical discourse.

Patek Philippe Grand Complication Ref. 5208P-001 (ca. 2011–2018, MSRP: approx. $1 million). 42 mm platinum case with concave bezel and hollowed-out lugs.

The movement consists of 701 parts, including 331 parts for the chronograph and 160 parts for the instantaneous perpetual calendar (changing over to the next day precisely at midnight—a rarity). Calibre R CH 27 PS QI incorporates fifty-eight jewels, although more recent iterations of the movement utilize sixty-three jewels and 719 parts.

The chronograph is a monopusher—the button at two o'clock is used for start, stop, and reset functions. The central second hand is for the chronograph.

The lever and escape wheel are in Silinvar (Patek Philippe's trademarked term for its silicon-based material).

The minute repeater activates with a tug of the lever on the left flank of the case and rings with classic gongs. Leap year indication is via an aperture within the subsecond dial at three o'clock (here indicating III—i.e., year three in a four-year leap cycle).

A very rare and special timepiece, it was discontinued in 2018, with only the rose gold version now offered for sale. Its exceptionally steep price point makes it accessible to only a few very fortunate individuals.

1. First known use of "horology" to describe the science of measuring time—1819: from Greek hōra "time" + English -logy (itself derived from French -logie or Medieval Latin -logia, or from Greek -logos). First known use of "logos" was in 1587, denoting the Divine Wisdom manifest in the creation and running of the world.

SUN
20
PATEK PHILIPPE
GENEVE
12
9
3
6
60
50
10
40
20
III
SWISS

If you want to jump straight into the topic of wristwatches, then you may wish to skip past the chapter on "The Measurement of Time" and the chapter that follows, "The Philosophy and Science of Time (and Space)." You can always revisit these sections later, when feeling more reflective and focused. The chapter on philosophy conveys complicated subject matter and requires deliberation and concentration if you plan to get the most out of it.

If you are so inclined, you will read the detailed footnotes. I wrestled internally regarding the incorporation of supplemental footers or whether to go with more-detailed and word-heavy body text sections. The former won out. Some topics are remarkably detailed or interesting—at least they are to me, and I wanted to share them with you without unnecessarily cluttering the textual flow. However, if you are not interested in shifting your eye line every few moments (and I certainly do not blame you—it can be nauseating to some), I doubt it will have a significant bearing on the comprehension of the remainder of the book. However, you may sacrifice some rather esoteric and thought-provoking content. If you are as captivated by minutiae as I am, then you will enjoy reading them, but do not let their placement spoil your flow. You can always come back to the footnotes.

It is this last point that I wish to stress. The book in your hands is my tribute to the vast universe of horology. Hundreds of minute[2] details are spread throughout its pages that I hope you come back to read. Without a doubt, if my wife had not pushed me to "publish it already," this digest would be several hundred pages longer and heavier. Perhaps, if there is a second edition, I will be able to share more horological matters with you.

I have included background about the watch industry and a discussion on the components of a watch. I have also examined the ascendant smartwatch industry, which seems to be exhibiting a sort of déjà vu for the traditional analog watch industry, almost decimated by the last shift in technology fifty years prior, when inexpensive quartz watches threatened mechanical watchmaking with extinction.

Every industry has its sordid side. This book deals mostly with the magical aspects of watchmaking, so I have kept my critiques to a minimum.

I have provided retail or market pricing (e.g., last known auction price or selling price) for almost all the watches under discussion, to offer some kind of relative scorecard for the watch. Price is not always (or even consistently) a good prognosticator of quality, durability, or desirability, and it is not my intention or desire to ascribe worthiness to a timepiece solely, or even mostly, on the basis of its price. Moreover, one need not spend exorbitantly to enjoy watch collecting as a hobby. However, I would be remiss for not pointing out how extraordinarily expensive watches may be—at times undeservingly so; the overall value presented in a timepiece is too often misaligned with its lofty price point.

I hope this book finds its way into the hands of many people and many cultures. Indeed, the world is a better place for having us all in it. Several authentic religious texts have been referenced, necessitating additional care to be taken to prevent the desecration of G-d's Hebrew/Arabic/Aramaic Name and "English Name," which not only should be carefully pronounced, according to Judaic, Christian, and Muslim mandate and tradition, but should also only be properly written in full form in religious texts—which this book is not.

I certainly hope that this title ends up as a reference on your bookshelf, but it may just as easily find its way into the trash heap. While I would prefer an alternative outcome, as you saw but a few sentences prior, to ameliorate this concern, the hyphen or dash "-" represents the "o" in G-d's "English Name" and for any Hebrew presentations of His Name, a Hebrew kuf "represents" a heh. אֱלֹקִים is apropos in Hebrew for writing His Name in a text that may end up being discarded (while still providing native speakers and adherents with original source terminology), sensitivity notwithstanding. Using the name Hashem, Allah, or G-d demonstrates equal humility among religious adherents of the aforementioned religions. Thus, all religious quotes will be as they appear in the source, but for the sensitivities noted.

Dear reader, I would be most gratified if after reading this book, I have fostered within you a newfound, renewed, or enhanced love of watchmaking and an increased appreciation of the value of time.

– Barry B. Kaplan

2. Interestingly, the word "minute" has numerous definitions and several pronunciations. True, it is the sixtieth part of an hour (among other definitions) when considered a noun, but as an adjective, and when pronounced "my-NEWT," it means very small (infinitesimal). Used as a transitive verb, it means to make notes or a brief summary of an event. It originates from the Late Latin minutus—made small.

INTRODUCTION

"The future is something which everyone reaches at the rate of sixty minutes an hour, whatever he does, whoever he is."

—C. S. Lewis

Consider the following very brief thought experiment: You awaken in an unfamiliar room, blinds tightly drawn over the windows, absolutely no external light penetrating the inky-black quiet darkness.

What are your first thoughts?

"Where am I?"

"What time is it?"

Our lives are for better or worse ruled by time's inexorable march. We make arrangements for ourselves and with others and use time as our guide for numerous starts and completions throughout the day and throughout our lives.

Seconds, minutes, hours, days, weeks, months, seasons, years, and decades pass by and accumulate—slowly for some, too rapidly for others. Modern psychology attributes time's *perceived* tempo to the unique characteristics and degree of prior life experience—the older we are, the quicker time appears to change for us. Truly novel life experiences accumulate more slowly as we age. We see this contrasted against the experience of young children, who seem to perceive time's advance as slow and deliberate, with intervals between birthdays and anniversaries feeling long and drawn out for them. Most adults perceive time whizzing by. Before one birthday passes, another appears in the headlights. One suggested antidote is to seek out new and diverse experiences as we age.

Understanding the significance and implications of time itself is a philosophical labyrinth. Philosophers have debated its meaning (and even its very existence) for millennia. I have a sensation of the passing of time, but precisely what am I sensing, and how am I sensing it? Classical philosophy presents time as comprising past, present, and future. With that representational template in mind, the past is immutably fixed and the future is at least partly undefined. With time's passing, present becomes past and future becomes present. Time elapses with a distinct present moment moving (as it were) toward the future. Presentism, contained within this basic intuitive understanding, argues that only the present exists. It does not travel forward; the present simply changes. We will discuss much more about the philosophy and science of time in a dedicated chapter.

My childhood bridged the late '70s and '80s, just as digital quartz achieved maturation. This epoch would later become known as the "quartz revolution." Analog watches waned in popularity as inexpensive quartz watches rapidly replaced them, many with add-on features such as calculators, video games, alarms, and timers. James Bond, in keeping up with the times, traded in his Rolex for a digital Hamilton Pulsar in 1973's *Live and Let Die*.

Then, in the mid-1980s, analog watches got their groove back. The Swatch watch took what some perceived as an otherwise boring analog watch and made it inexpensive and cool. Because Swatch had fully automated the wristwatch-manufacturing process, the cost of production declined 80 percent and the number of required parts reduced by almost half, while still maintaining engineering tolerances and timing accuracy.[3]

3. Elmar Mock, coinventor of the Swatch watch: "Our big innovation was to use ultrasonic welding to build the mechanism straight into the case, which was made of the same type of plastic used in Lego. We decided that every screw was a danger: if you can screw something, it can unscrew. So everything is welded, making it watertight. Since this means it cannot be taken apart, we had to produce something of the highest quality that would not need repairing. It also meant we could halve the number of parts to just 51. The result was a watch three times cheaper than any other that could be produced in Switzerland." *The Guardian*, January 8, 2018.

Originally intended to recapture lost market share ascribed to the introduction of (inexpensive) quartz watches—by repopularizing the analog watch—these colorful and artistic timepieces became a global sensation.[4] People would apparently return to store openings in disguise to bypass restrictive stock allocations. By Swatch's fifth anniversary (1988), it had produced its fifty millionth watch. By 1992, Swatch had shipped one hundred million units. That number had grown to three hundred million by 2003, and at the end of 2017, it was seven hundred million.[5]

Some watch industry historians consider Swatch's success to have (almost) single-handedly saved the Swiss watch industry. This success afforded the Swiss sufficient time and resources to merge ailing businesses and successfully reposition the Swiss watch as an enduring luxury item *to be passed on to future generations*.[6]

Modern smartwatches and smartphones have built-in obsolescence. Technology catapults us all ahead in an unceasing and dizzying cycle, with transistors constantly miniaturizing, processing power growing exponentially, battery life continually extending, and screens relentlessly enlarging and increasing in brightness and pixel density. Even the most-powerful modern devices have a useful life of no more than five years without necessitating a serious upgrade, which in most cases is infeasible. The original Apple Watch began preorders on April 10, 2015. The Apple Watch Series 6 became available from September 18, 2020, onward. In the short span of five years, the product has gone through six product development cycles and has become many times more powerful and capable than its progenitor. Case in point regarding built-in obsolescence: after just five years, the original Apple Watch will not run the latest Apple Watch operating system.

A magnificent timepiece is a work of *mechanically engineered art*, easily appreciated by the uninitiated and serviceable today with much the same techniques as two centuries ago. A mechanical watch, properly maintained, will last almost forever. Therein lies the appeal. It records time's eternal march—and becomes a part of it.

A rare Panerai Radiomir Ref. 3646 tropical "Brevettato" (ca. 1940). Case and movement by Rolex. The dial was originally black, but with eighty years of exposure to the elements (and a minor degree of radioactivity—see below), it has developed a deep-caramel-brown patina.

Panerai continues to produce the instantly recognizable cushion-shaped Radiomir in a variety of lug and crown styles, including an almost identical modern version of the illustrated timepiece.

The name "Radiomir" is suggestive of the proprietary radium-based powder historically used to illuminate Panerai's dials. Indeed, the concentric circular pattern barely visible toward the dial's center is most likely the result of many years of low-level radiation reflecting off the glass and discoloring the dial.

4. In 2015, Sotheby's in Hong Kong sold the "Dunkel collection" for $6 million. The collection comprises 5,800 Swatch watches (and prototypes) spanning twenty-five years.

5. *The Guardian*, January 8, 2018.

6. With apologies to Patek Philippe's enduring Generations campaign, "*You never actually own a Patek Philippe. You merely look after it for the next generation.*" Through perhaps the most brilliant advertising campaign ever conceived for luxury goods (after "*A diamond is forever*"), Patek has justifiably earned the right to proclaim its timepieces as heirlooms. The brilliantly crafted tagline, running since the late '90s, targets the aspirational consumer, the so-called emerging affluent.

Patek Philippe's marketing genius is in its ability both to interpret and transform cultural norms. Depictions of men with their sons in an office, on a yacht, or in a luxury automobile portray the ultimate presumed aspirational lifestyle. If you desire to live this lifestyle, then by extension, you should aspire to own this timepiece. Judging by the price appreciation of almost every complicated Patek Philippe model in the preceding two decades, it has also proven (at least so far) to be a remarkably resilient long-term investment.

When the Great Recession hit, Patek Philippe—seeking to mitigate demand disruption—desired additional wrists. Thus appeared mother and daughter to "*Begin your own tradition*"—"*Something truly precious holds its beauty forever.*" These judiciously art-directed advertisements ensure picture-perfect poses of mother and daughter in an idealized (almost dreamlike) world of the archetypal mother-daughter relationship—at least as some would define it. This ad appears to target female consumers, but it not so subtly targets gift-seeking men. In magazine ads from the first few years of the 2000s, Patek Philippe portrayed young elegantly attired women alongside the tagline "*Who will you be in the next 24 hours?*" In some countries, the women wear wedding rings; in others they do not, or at least it is impossible to know. Undeniably, Patek Philippe's image-marketing policy subtly aligns with those territories' accepted cultural norms in respect to marriage infidelity.

RADIOMIR
PANERAI

THE MEASUREMENT OF TIME

"Better three hours too soon than a minute too late."

—William Shakespeare, *The Merry Wives of Windsor*

Throughout history, the sun, moon, planets, and stars have provided us with a means for measuring the passage of time and determining the seasons, months, and years. Little survives concerning prehistoric timekeeping, but ancient artifacts indicate that people of all civilizations were to some degree preoccupied with measuring and recording time's passage.

Twenty-thousand-year-old architectural remnants[7] discovered in Europe exhibit scratched lines and gouged holes in sticks and bones, possibly used by prehistoric humans to count the days between moon phases. Ancient civilizations also utilized the day and the solar year to keep track of time, but the lunisolar[8] calendar was antiquity's most popular method for tracking and synchronizing extended periods.

Our current sexagesimal (or base 60[9] numeric system) of time measurement is commonly ascribed to the Sumerians ca. 2000 BCE. We continue to use a modified form of this system today because of its ease of use in measuring angles, geographical coordinates, and time. Base 60 gives us sixty-minute hours and sixty-second minutes.[10]

The Egyptians utilized base 12 (duodecimal) as their system of counting. Using one's thumb as the counting "pointer," one would count the three joints on each finger, yielding twelve joints. As with sixty (and as explained in a preceding footnote), twelve is also a superior, highly composite number and has six factors, making it practical for denominating hours. Ancient Egyptians split the day into two equal periods of twelve hours. However, these hours were variable and would differ with the changing seasons—thus, hours in

In 2013, during archeological excavations in the Kings' Valley in Upper Egypt, near what were once stone huts housing gravesite construction workers, University of Basel researchers unearthed one of the world's oldest Egyptian sundials dating to ca. thirteenth century BCE.

Led by Professor Susanne Bickel, the team made the noteworthy discovery while clearing the entrance to one of the tombs. The flattened limestone ostracon[11] features a semicircle in black divided into twelve sections of about 15 degrees each.

A hole is located in the middle of the nearly 17-centimeter-long horizontal baseline. A rod inserts into this hole and casts a shadow displaying daylight hours. Small dots (located roughly three-quarters toward the middle-top of each segment) were used for even more detailed time measurements.
Image source: University of Basel, public domain

7. Various estimates range from 10,000 BCE to as far back as 30,000 BCE.
8. A calendar where the months are lunar but years are solar, synchronizing almost exactly every nineteen years.
9. The number 60 is a "superior highly composite" number (the fourth one after 2, 6, and 12). Basically, a superior highly composite number has more factors than any other natural number before it AND on a weighted basis is superior to any other natural number larger than it in terms of the divisibility of the number by its factor count. Whew! Simply put, it means that 60 can be divided (without leaving a remainder), by the numbers 1, 2, 3, 4, 5, 6, 10, 12, 15, 20, 30, and 60. Thus, twelve factors in all, including three primes: 2, 3, and 5. By having so many factors, the sexagesimal numbers are easily simplified. An hour, for example, can divide into equal parts consisting of thirty minutes, twenty minutes, fifteen minutes, twelve minutes, ten minutes, six minutes, five minutes, four minutes, three minutes, two minutes, and one minute. Base 60 is therefore an exceptionally practical and useful method for counting seconds and minutes.
10. A degree of arc divides into 60 minutes, and each minute into 60 seconds. In astronomy, the arcminute and arcsecond have been used since antiquity: in the ecliptic coordinate system as latitude (β) and longitude (λ), in the horizon system as altitude (Alt) and azimuth (Az), and in the equatorial coordinate system as declination (δ). All are measured in degrees, arcminutes, and arcseconds. A degree is 1 part in 360 of a turn; an arcminute is 1 part in 60 of a degree; and an arcsecond is 1 part in 60 of an arcminute or 1/3,600 of a degree.
11. A potsherd (piece of pottery) used as a writing surface.

1 2 3 4

winter would be "shorter" than hours in summer. Archeologists have uncovered 3,500-year-old sundials segmented into twelve sections, each 15 degrees apart (totaling 180 degrees), to account for the sun's apparent daily east–west hemispheric rotation.

Thirty-six decans make up the twelve familiar constellations. A decan[12] is each of three equal 10-degree divisions of a sign of the zodiac. Every (approximately) ten days, a new decanic star group reappears in the eastern sky at dawn (just prior to sunrise). Thus, 12 zodiacal signs × 3 (divisions) × 10 (days per division or approximate degrees per division) = 360. The ten-day approximation leaves room for the remaining (approximately) five days to make up the Egyptian Sothic cycle (365.25 days)—equivalent to the reformed Julian calendar (more on that later). The Sothic cycle shares similarities with the sidereal year (365.2564 days, which is based on the earth's rotation around the sun relative to the position of fixed stars) and the solar (or tropical) year (365.24217 mean solar days, which is the time it takes for the sun to appear in the same position in the sky as seen from the earth).

During the night, twelve decans rise, each rising decan representing the beginning of a new decanal hour. From as early as 2100 BCE, ancient Egyptians utilized this method for nighttime celestial time measurements.

The Egyptians also utilized large obelisks to track the passage of the sun. Subsequent clocks included water clocks[13] (especially useful during the night, when the sun was not visible—and one of antiquity's most accurate timekeeping devices), candle clocks,[14] timesticks,[15] hourglasses,[16] and sundials.[17]

EARLY CALENDARS

More than four thousand years ago, the Babylonians (in present-day Iraq) implemented a year of twelve alternating twenty-nine-day and thirty-day lunar months, giving rise to a 354-day year.

Sumerian scribes used a 360-day year (12 × 30) as early as 2400 BCE.[18] As an approximation of the combined average (359.8048 days) of the lunar year (354.3671 days) and solar year (365.2425 days as expressed by the Gregorian calendar), the 360-day calendar differs by just 0.1952 days and was thus a useful proxy for either calendric system.

French pillar dial, ca. 1600–1800, measuring 76 by 22 mm. Hand outline shown to indicate scale. The pocket-size device mirrors the dimensions of a tubular lipstick container.

The gnomon is stored inside the device when not in use. The top bell-shaped section is rotated to move the gnomon to the designated month (indicated at the base) in order to properly discern the approximate hour for that time of year.

12. The origin of the word "decan" is the Greek word dekanoi, meaning tenths.

13. A water clock or clepsydra is a timepiece that measures time via a regulated inflow or outflow of a liquid from or to a vessel. Numerous ancient civilizations from the Chinese to the Egyptians utilized these devices for time measurement.

14. Candle clocks consist of thin candles with a precisely measured mass of wax, uniform thickness, and spaced divisions marked either on the candle or on the device in which it stands. To prevent accelerated burning from wind or other atmospheric effects, a shielded casing protects the candles.

15. Also known as a shepherd's dial, shepherd's stick, pillar dial, or cylinder. It is a type of sundial, but it is portable and consists of a handheld wooden cylinder. Named for the shepherds who would score their staffs with a monthly scale that measured the altitude of the sun for a particular latitude and time of year, it features a movable protuberance (the gnomon) that juts out from the side of the cylinder and extends toward the sun to cast the necessary shadow.

16. An hourglass, sandglass, sand clock, egg timer, or sand timer measures the passage of time through a regulated flow of a substance (usually sand) from one bulb (the upper one) to another (the lower one) in a two-bulb vertical configuration. The bulbs are usually symmetrical to allow for time measurement regardless of orientation. Adjusting the size of the bulb, the coarseness of the granular-flow particulate, and the width of the neck connecting the two bulbs regulates time measurement.

17. Probably the best-known ancient time-telling device, the sundial measures the altitude or azimuth (angular distance) of the sun by using a gnomon (usually a wedge-shaped protuberance) that casts a shadow onto a dial, which is frequently a flat plane (but can be various other shapes).

18. *Encyclopedia Britannica*, "Ancient and Religious Calendar Systems":britannica.com/science/calendar/Ancient-and-religious-calendar-systems.

Shadow clock (ca. 306–30 BCE). Purchased from the Egyptian government in 1912 by New York's Metropolitan Museum of Art, this "portable" timepiece dates to the Ptolemaic period and was unearthed in the Theban necropolis by Howard Carter and Lord Carnarvon. With a mass of 1.3 kg (2.9 lb.), it weighs about the same as a modern laptop computer. It measures 9.3 by 5.2 by 14.7 cm (height by width by depth)—about the same cubic volume as nine decks of cards.

As presented, the front of the base, just beneath the sloping face, is broken off. Were it complete, the base would have extended forward. A perpendicular block positioned in front of it acted as the gnomon, casting the necessary shadow on the sloping face. Gouged with six radiating lines, each corresponding to two months of the year, the sloping face indicates the hour calculated by looking at where the sun's shadow intersects with both the line corresponding to the relevant month and the oblique line corresponding to the relevant hour.

The mesmerizing Ikepod Steel Nanoball Timer/Hourglass (ca. 2011, MSRP: $28,500–$40,000 for sixty-minute versions, $13,000–$17,500 for ten-minute versions). Instead of sand, this modern reimagining of the traditional sandglass utilizes eight million 0.3 mm steel nanoballs (or 1.3 million nanoballs for the ten-minute version) housed in a single piece of industrial borosilicate glass.

Conceived by the acclaimed designer Marc Newson, it is available in four nanoball colors: black stainless steel, stainless steel, gold, or copper. The larger version measures 300 mm tall by 250 mm wide and weighs a mammoth 9.6 kg.

Early Egyptian calendars followed lunar cycles. Later, however, they realized that the bright star we now call Sirius[19] rose next to the sun every 365 days. Thus, according to historians, almost five thousand years ago,[20] they devised a 365-day solar calendar.[21]

The Mayans of Central America, however, relied not only on the sun and moon, but also on the planet Venus to establish two calendars—one of 260 days, and another of 365 days.

The Hebrew calendar traces its origins to the teachings of Shem (son of Noaḥ, born ten generations after Adam; see later footnotes from the *Kuzari*), who according to biblical accounts lived from 2203 to 1603 BCE.[22]

Copious research abounds regarding the aforementioned Sumerian, Egyptian, Mayan, and Babylonian calendars. There is a relative dearth of information elucidating the Judaic calendar. This is the primary reason why I have focused on examining the Hebrew/Jewish calendar more closely than the calendars listed above. Much of what follows below regarding Hebrew calendar computations would equally apply to the understanding of the previously mentioned calendric systems.

THE HEBREW/JEWISH CALENDAR

There is significant debate on the origins of the lunisolar calendar. While many modern scholars insist that the Hebraic lunisolar calendar dates to around 600 BCE,[23] extensive biblical sources and contemporaneous writings conflict with this assertion. The discussions below are excerpted from several writings encompassing many sources (spanning several millennia) and referencing the Jewish written and oral tradition[24] leading as far back as Abraham,[25] his progenitor Shem, and his ancestor Adam.

The following page shows one of the most enigmatic devises ever discovered. Displayed at right are the front and back of the largest of eighty-two extant fragments of what was once an astronomical computer. To quote Dr. Brenda Wyatt, the forensic scientist/detective from 1986's *Highlander*,[26] "It would be like finding a 747 a thousand years before the Wright brothers ever flew."

Scientists have dubbed it the Antikythera (an-tiki-th'ear-a) mechanism. Seven large fragments and seventy-five small fragments were recovered from a Roman shipwreck lying between Crete and the Peloponnesus on the edge of the Aegean Sea near the island of Antikythera, following a spring 1900–1901 sponge-diving expedition off the island's east coast.

Dated to approximately 70 BCE (more than 2,000 years ago), this curious 20 centimeter-tall instrument was discovered by the divers, among other rare archeological finds from the ancient shipwreck. Now preserved in the National Archaeological Museum of Athens, it demonstrates that **at least two millennia ago**, Greek civilization was sufficiently sophisticated to design and construct a brass astronomical machine of exceptional complexity, refinement, and functionality.

19. Sopdet in ancient Egyptian, and Sothis in Greek.
20. H. E. Winlock, *The Origin of the Ancient Egyptian Calendar* (1940). Winlock writes, "Alexander Scharff of the University of Munich had long seen the difficulties inherent in Meyer's theory, and in 1927, he had stated that the calendar must have been invented in 2773 B.C.—a whole Sothic period later than had usually been proposed. Before that date he assumed that the Egyptian reckoned time by some wholly different system, which he did not exactly define but which in one place he seems to say was based on a year of 320 days."
21. The Gregorian calendar (named after Pope Gregory XIII, who introduced it in October 1582) is the widely accepted universal standard and is an example of a solar calendar. It is based on the Julian calendar (adopted by Julius Caesar approximately 46 BCE), which in turn is based on the Egyptian calendar.
22. Shem's lineage according to the Hebrew Bible: Arpaḥshad, Shelaḥ, Eber (with whom he established a seminary), Peleg, Reu, Serug, Naḥor, Teraḥ, Abraham.
23. *Encyclopedia Britannica*, "Ancient and Religious Calendar Systems": britannica.com/science/calendar/Ancient-and-religious-calendar-systems.
24. This point is pertinent—the Oral Law in Judaism is a fundamental component of Torah. Without it, the Written Law would be frequently indecipherable—for example, the construction of phylacteries ("tefillin"—a set of small, opaque, black leather boxes containing scrolls of parchment inscribed with verses from the Torah—one worn on the forehead, the other on the upper arm) would be a complete mystery without an oral tradition passed down from Moses. The Torah mentions only that specific passages should be written in the phylacteries and where they are to be worn, but says nothing about their design, fabrication, and assembly.

Until the Common Era, the oral component of Jewish law passed generationally (and unerringly) from teacher to student. It was the primary and sometimes singular method of instruction. However, numerous pogroms and resettlements of the Jewish community throughout the ages necessitated a lasting solution to ensure the Oral Law's continued integrity and permanence. Judah the Prince, completing a project instituted by his father (in collaboration with Rabbi Natan, a.k.a. Nathan the Babylonian), consolidated the various opinions into a single body of law known as the Mishnah. The Talmud (both Babylonian and Jerusalem versions) was compiled several hundred years later to further elucidate the deliberately condensed language of the Mishnah.

Judah the Prince: Also called Judah ha-Nasi or Yehudah HaNasi (in transliterated Hebrew)

Judah the Prince was a second-century CE rabbi and editor of the Mishna (a written collection of the Oral Law). The title "nasi" was used for presidents of the Sanhedrin (an assembly of seventy-one elders appointed to institute and discharge legal duties in accordance with Torah principles). He was the first nasi to have the title permanently appended to his name. His teachers and contemporaries were some of the greatest luminaries in Talmudic literature. The Mishnah refers to him simply as Rebbi, Rabbeinu, or Rabbeinu HaKadosh (our holy Master).
25. Abraham is the patriarch of Judaism, Islam, and Christianity.
26. *Highlander* is a cult film about immortal beings engaged in battle through the millennia, decapitating each other until only a single immortal remains.

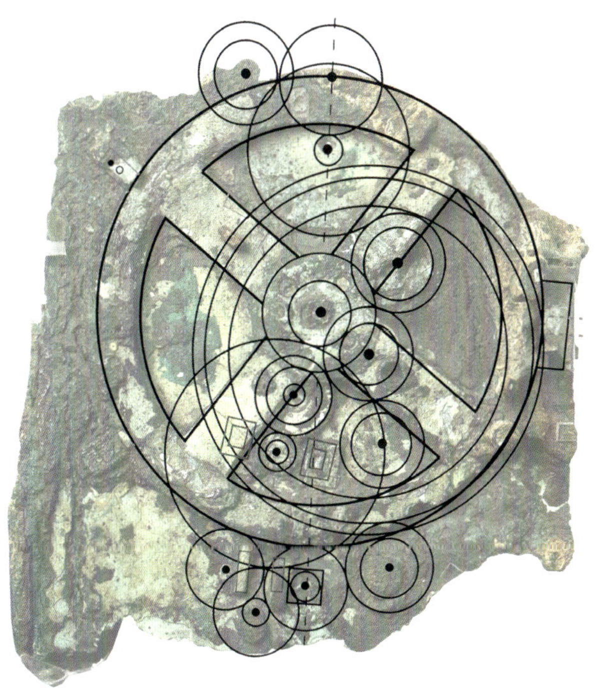

Although discovered at the very beginning of the twentieth century, and while it was immediately determined to be an astronomical machine of some kind, it took another century before researchers could peer beneath the fragile surface to look in more detail at its inner mechanism. Using twenty-first-century x-ray computed tomography and advanced 3-D imaging (which allows cross-sectional slices of the object's internals to be individually imaged), researchers discovered that the device contained numerous gears and inscriptions. The obscured inscriptions provide instructions for the device's operation.

Combining these instructions with known astronomical data and knowledge about the toothed gears hidden within, researchers were able to reconstitute the mechanism's purpose, functionality, and period of manufacture. In fact, on the basis of the accumulated research data, several engineering-oriented hobbyists have produced modern re-creations.

Both faces of the device display information. The front face featured a large dial exhibiting a 365-day solar calendar. Inscribed within this dial were the twelve signs of the zodiac. Researchers determined that by turning a crank on the side of the device, the dial would have indicated the exact position of the sun, moon, moon phase, and zodiac for every day of the Egyptian calendar year. The device's rear upper face displayed the nineteen-year Metonic cycle (235 lunar months) overlapping with the solar calendar. Below the top spiral-shaped dial lay a second, similar dial that represented both solar and lunar eclipses. Inspired by the Antikythera mechanism, Mathias Buttet, Hublot's director of research and design, re-created the original device in a modern 30 mm movement housed in twenty limited-edition timepieces. The Hublot Antikythera SunMoon consists of 295 parts and seven complications, including a solar calendar, lunar calendar, and sidereal position of the sun and the moon.

No similar ancient device is currently known to exist, but it is conjectured that this and other astronomical devices were commissioned for the cultural and scientific elites of the era. Perhaps lying in wait for discovery by future archeologists are even more interesting and sophisticated ancient computational machines.

The image at top left overlays the gear-train configuration to provide a more complete representation of the mechanism's layout.

ORIGINS OF THE JEWISH CALENDAR

The Babylonian Talmud[27] in tractate (volume) Rosh Hashanah[28] gives the average moon phase duration as twenty-nine days, twelve hours, and 793/1080 parts of an hour.[29] The eleventh-century Talmudist Ḥananel ben Ḥushiel notes that this tradition originated with Moses.[30] Modern cosmologists denote the average moon phase as twenty-nine days, twelve hours, forty-four minutes, and 2.8016 seconds—**a mere half-second deviation** per lunar month from the values disseminated by Jewish tradition.[31]

Such precision was formerly ascribed to Ptolemy via his work titled *Almagest* (the Great Book).[32] Ptolemy cites the renowned Greek astronomer and mathematician Hipparchus as his source (ca. 190–ca. 120 BCE). Historians studying Hipparchus's works indicate that he used Babylonian sources extensively in his astronomical calculations. Were these Babylonian sources formerly Hebraic? After all, at least since the time of Moses,[33] the Jewish people have understood how to calculate the moon's synodic period.

As the Torah[34] relates, the first commandment given to the Jewish people (in 1313 BCE, prior to their departure from Egypt) was that of marking and celebrating the new moon festival (or Rosh Ḥodesh) and establishing a calendar henceforth. It was of primary spiritual importance to properly calculate and ascertain the occurrence of the new moon in order to facilitate the timing of religious observances and festivals.

The Hebrew calendar is one of many extant lunisolar calendars.[35] It necessitates a lunisolar arrangement in order for the festival of Passover to maintain celebration at its proper time (i.e., in the spring).[36] When required, an added lunar leap month prevents Passover from occurring in the winter.[37] The two calendars almost perfectly coincide every nineteen years.[38]

Such a calendar indicates both the moon phase and time of the solar year.[39] It should be noted, however, that while calculations of days, months, and years utilize *fixed hours* equal to one-twenty-fourth of a day[40] (each hour segmented into 1,080 subsections equivalent to 3⅓

27. The Talmud is Judaism's central rabbinic text and the primary source of Jewish law and theology. The Babylonian Talmud consists of sixty-three tractates (volumes) and 2,711 double-sided pages. It comprises the Mishna (the oral law as handed down by Moses ca. 1393–1273 BCE and compiled ca. 200) and the Gemara (rabbinical analysis and commentary on the Mishna), compiled ca. 500.

28. Meaning "head of the year."

29. 25A: "Rabbi Gamliel said to them: This is the tradition that I received from the house of my father's father: The monthly cycle of the renewal of the Moon takes no less than twenty-nine and a half days, plus two-thirds of an hour, plus seventy-three of the 1,080 subsections of an hour."

30. Ḥananel ben Ḥushiel, born ca. 990, comments on Exodus 12:2: "We have an ancient tradition that for the twelve months of the year, five have thirty days each, whereas another five have twenty-nine days each. The remaining two months sometimes have twenty-nine days and other times thirty days. Sometimes, one of the two months in question has thirty days, whereas the other one has twenty-nine days. The two months referred to are always Marḥeshvan and Kislev. We also have an ancient tradition that the first day of the month of Tishrei is at the same time as the first day of the New Year. **Every single month, according to tradition handed down since the time of Moses, consists (astronomically speaking) of 29 days 12 hours and 793 parts of the 1,080 parts into which 1 hour (60 minutes) is divided in Jewish law."**

31. In seconds, the traditional Hebraic value is 29 × 86,400 (seconds per day) + 43,200 (seconds in 12 hours) + 793/1,080 × 3,600 (seconds in one hour) = 2,551,443.333 . . . seconds. The modern cosmologically derived value is 2,551,442.8016 seconds, a deviation of only 0.53173 seconds or about half a second.

32. Otto E. Neugebauer's *A History of Ancient Mathematical Astronomy* (1975), page 69: "Ptolemy inherited from Hipparchus the following parameters; mean synodic month: 29;31,50,8,20^{d}." The *Almagest* is a second-century Greek-language mathematical and astronomical discourse regarding the apparent motions of the stars and planetary paths. Written by Claudius Ptolemy (ca. 100–ca. 170), it is considered by many to be one of the most influential scientific texts ever written.

33. Moses (1393–1273 BCE) is the most important Jewish prophet and, following Jewish custom, the greatest prophet to ever live. According to the Hebrew Bible, G-d deployed him to liberate the Israelites from crippling servitude in Egypt. With G-d's Divine assistance, he freed the Israelites, visited ten plagues upon the Egyptians and Egypt, and over the course of forty years faithfully led the Israelites through the desert to the land of Israel, while performing miracles and wonders. At Mount Sinai, after G-d spoke the Ten Commandments, Moses transmitted the laws (**both Written and Oral**) to the people.

34. Torah has a range of meanings, from the traditional five books of Moses (Pentateuch or Ḥumash) to the cumulative knowledge of Jewish teaching, culture, and practice. Here the term is referring to the five books of Moses.

35. Babylonian, Buddhist, Chinese, ancient Hellenic, Hindu, Japanese, Kurdish, ancient Mesopotamian, Mongolian, Tibetan, and Vietnamese calendars, among others, are all lunisolar.

36. Deuteronomy 16:1: "**Keep the month of spring** and make the Passover offering to the Lord, your G-d, for in the month of spring, the Lord, your G-d, brought you out of Egypt at night."

37. Babylonian Talmud, tractate Rosh Hashanah, 21A.

38. Also known as the Metonic cycle, this nineteen-year cycle consists of 235 synodic lunar months. In this cycle, years 3, 6, 8, 11, 14, 17, and 19 all contain a lunar leap month. Which seven of the nineteen years to use was not conclusively determined until the time of Saadia Gaon (ca. 882–942 CE). For the mathematically inclined, 235 synodic months (lunar phases) = 6,939.688 days (the Metonic period). Nineteen tropical years = 6,939.602 days. This would result in a single-day difference every 219 years.

39. Lunisolar calendars, which divide the year into months, require that the year have a whole number of months. Ordinary years consist of twelve months, but every second or third year is an embolismic year, which adds a thirteenth intercalary, embolismic, or leap month.

40. Equal or equinoctial hours are one-twenty-fourth of the day as measured from noon to noon; the minor seasonal variations of this unit eventually smoothed by making it one-twenty-fourth of the mean solar day.

seconds[41]), the beginning of each halaḥic[42] day utilizes local sunset and is thus variable on the basis of an observer's geographical location.

Genesis 1:14[43] describes different time units determined by the relationships between the celestial luminaries, including seasons, days, and years. Then in Exodus 12:2, the Torah recounts: "This month shall mark for you the beginning of the months; it shall be the first of the months of the year for you."[44]

Incorporating the aforementioned sources with his extensive astronomical knowledge, Rabbi David Shapero[45] provides an unusual and seemingly unique observation for the origin of the twenty-four-hour day:

> *It [Torah] further instructs that months be determined by the reappearance of the Moon after lapping the Sun every 30 days, as they both sweep west to east through the 12 mazalos*[46] *at different rates. Both the Sun and the Moon subtend an arc of approximately 1/2 degree, measurable as 2 minutes both rising and setting.*[47] *The Sun takes nearly 365.25 days to circle 360 degrees, very close to 1 degree per day, [or] two Sun diameters per day. Imagining the Sun as a sweep second hand on the cosmic dial of 360 degrees, it moves one [solar] diameter by day and one by night. But the Moon catches the Sun every 30 days, so 360 degrees times 2 lunar diameters = 720 lunar diameters / month. 720/30 = 24 per day. So using the Moon as a sweep second hand through 360 degrees*[48] *monthly yields 24 lunar diameters daily.*[49]

Abraham ben[50] Meir Ibn Ezra (ca. 1089–ca. 1164) and his eleventh-century Eurasian contemporaries were exceptionally knowledgeable regarding the twenty-four-hour day and the unavoidable distortion inherent in utilizing equal-length (fixed) hours. The great biblical commentator and philosopher of the Middle Ages (known colloquially as "the Ibn Ezra") authored numerous scientific works, including books regarding the use of the astrolabe.[51]

In "The Sabbath Epistle," he writes, "Anyone who says that the season is based on unequal (seasonal) hours, which are twelve in the day and also twelve in the night,[52] is more hopeless than a fool. For how is it possible that an arc that measures 105° at the equator should be like an arc that measures 255°?"[53]

ABRAHAM AND THE BABYLONIANS

Abraham's birth precedes that of Moses by more than 420 years. The Babylonian exile, a period in Jewish history during which many people from the ancient Kingdom of Judah were exiled to Babylon, occurred in 586 BCE.

Abraham,[54] the undisputed progenitor of Jewish, Muslim, and Christian faiths, spent his formative years in Ur Kaśdim (present-day Iraq). The territory lay adjacent to what would become the future seat of the Babylonian capital. Several historians proffer that the Babylonian

41. 1,080 subsections × 3⅓ seconds = 3,600 seconds = 60 minutes = 1 hour.
42. Halaḥic: relating to or connected with the Halaḥa—Jewish scriptural law and associated canonical texts and interpretations.
43. וַיֹּאמֶר אֱלֹקִים יְהִי מְאֹרֹת בִּרְקִיעַ הַשָּׁמַיִם לְהַבְדִּיל בֵּין הַיּוֹם וּבֵין הַלָּיְלָה וְהָיוּ לְאֹתֹת וּלְמוֹעֲדִים וּלְיָמִים וְשָׁנִים; G-d said, "Let there be lights in the expanse of the sky to separate day from night; they shall serve as signs for the set times—the days and the years."
44. הַחֹדֶשׁ הַזֶּה לָכֶם רֹאשׁ חֳדָשִׁים רִאשׁוֹן הוּא לָכֶם לְחָדְשֵׁי הַשָּׁנָה
45. Executive director Ohr Somayach, Detroit.
46. The twelve major astrological constellations.
47. We measure the day as 24 hours × 60 minutes = 1,440 minutes. Two minutes = 1/720th of a 360-degree circle, or ½ a degree.
48. Rabbi David Shapero: Technically when the moon laps the sun, it is more than 360 degrees, but from the perception of the observer, using the position of the sun as the beginning and the end point, it appears for the casual observer to be 360 degrees.
49. Rabbi David Shapero: This resonates with a line from the Yotzerot, Parashat HaChodesh (Eleazar ben Kalir, (ca. 570–ca. 640). קץ מולדתו חל להקצות, ביום רביעי בחצות, ועד שלושים מרוצות, לא ניכר בחוצות; where thirty hours are described as merutzot or "dashes of the Moon."
50. Ben (Bar in Aramaic, Ibn in Arabic) is Hebrew for "son of" and is part of a surname. In the case of Abraham ben Meir Ibn Ezra, it would equate to "Abraham, son of Meir, son of Ezra."
51. An astrolabe, literally meaning "star taker," is an analog device capable of solving numerous astronomical calculations.
52. Samuel ben Meir (ca. 1085–ca. 1158 CE) comments on Genesis 1:4: "He arranged for the day to be divided into twelve hours of daylight and twelve hours of night."
53. Ancient astronomers/astrologers used equinoctial hours just as we do. These are equal-length day and night hours as occurring at the equinoxes. Summer solstice in England extends daytime to seventeen equinoctial hours—an arc of 255°. At winter solstice, daytime is only seven equinoctial hours, an arc of 105°. Ibn Ezra found it unreasonable that an arc of 105° and an arc of 255° should divide into the same number of parts (12). The Sabbath epistle of Ibn Ezra, KTAV Publishing House, 2009.
54. The biblical account puts Abraham's birth at 1948 AM. Anno Mundi (Latin for "in the year of the world") abbreviated as AM. It is a calendar based on the biblical accounts of the creation of the world and subsequent history. Abraham lived from ca. 1813 to ca. 1638 BCE.

Astrolabe of the Rasulid sultan al-Ashraf (ca. 1291). Fashioned in brass and inlaid with silver, it measures approximately 8 by 6 by ¼ inches.

A stele (pronounced steely or steel) is an upright stone slab or pillar with a commemorative inscription or relief.

Shown here is perhaps the most illustrious stele of all, the Law Code of King Hammurabi, 1792–1750 BCE, fashioned in basalt, measuring 225 by 65 cm, and on display in the Louvre, Paris. This image portrays the back view of the famous object. On the front side in the top quarter, two figures are carved in relief: the Babylonian sun god and Hammurabi, reputedly receiving the law orally.

The circled region, enlarged here, is to show detail.

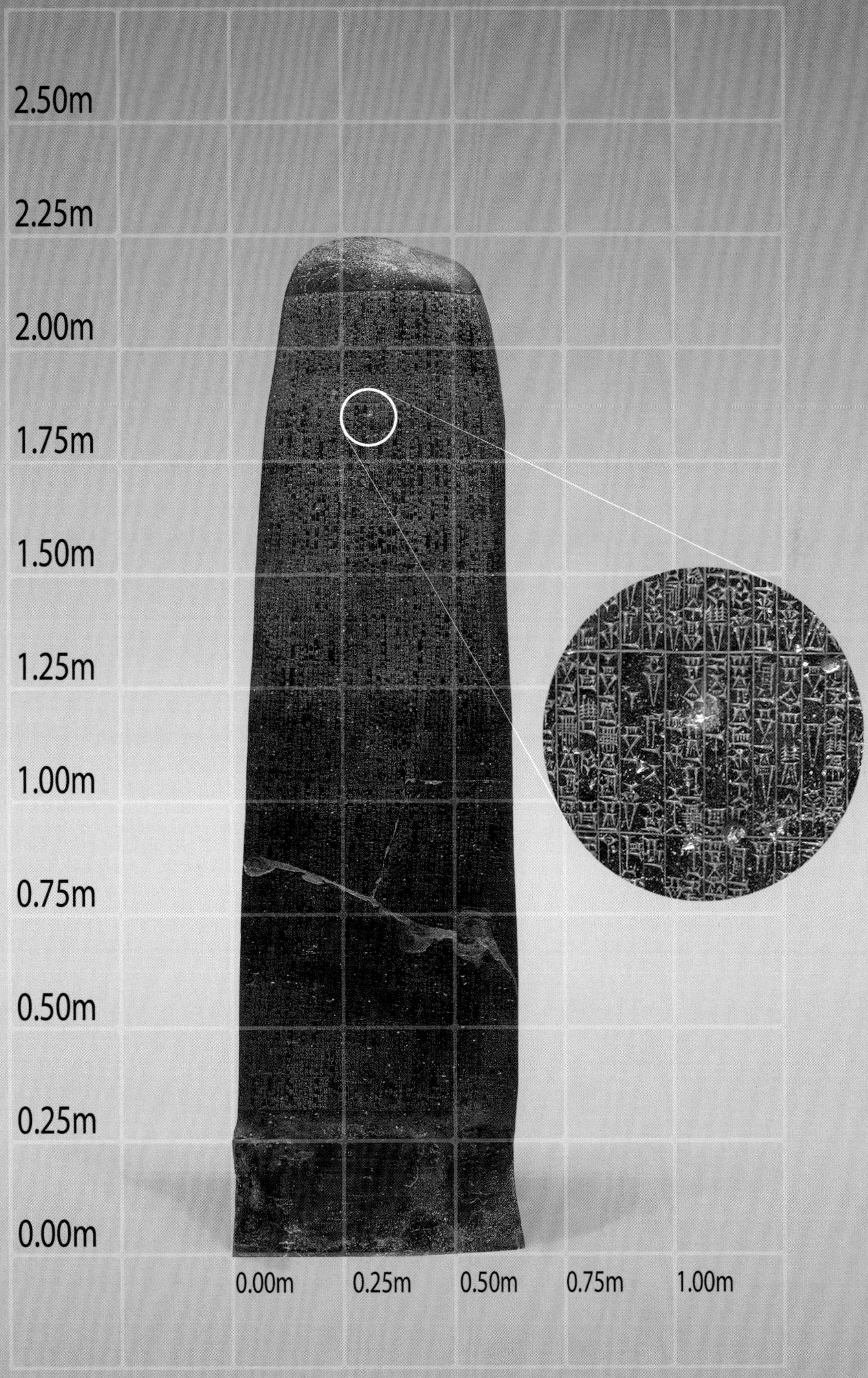

2.50m

2.25m

2.00m

1.75m

1.50m

1.25m

1.00m

0.75m

0.50m

0.25m

0.00m

0.00m 0.25m 0.50m 0.75m 1.00m

lunisolar calendar predates the Hebrew lunisolar calendar because the names given to the months by the Hebrews follow Babylonian monikers, such as Nisan or Adar. The convincing counterargument is that prior to the Babylonian exile, the Hebrews did not utilize names for the months at all (they used ordinal numbers instead: first, second, third, etc.) and thus adopted their oppressors' nomenclature when referring to the months by name.

Furthermore, Talmudic sources show that the tradition of intercalation[55] was well known and performed in Babylon, ***despite a general rule*** against the practice outside Israel.[56]

Modern secular scholars seem unfamiliar with the Talmudic directive that the intercalation not take place outside Israel, leading to mistaken assumptions that the Jewish calendar adopted Babylonian custom.

In Genesis 18, three "men," whom many commentators describe as angels, visit Abraham.[57] One of the angels (the archangel Michael) informs Abraham, "I will return to you next year, and your wife Sarah shall have a son!"[58]

Rashi[59] remarks that this event occurred during Passover.[60] Abraham would therefore have been well acquainted with the lunar month and its intercalation. His recognized religious piety in celebration of festivals and other observances would necessitate such devotion. Moreover, in order to calculate the eight-day festival of Passover, Abraham must have been exceptionally well acquainted with the workings of the lunisolar calendar and its nineteen-year cycle.

Several contemporaneous texts posit that Hammurabi,[61] the sixth king of the First Babylonian dynasty, was at least partially responsible for our modern lunisolar calendar. Hammurabi[62] and Abraham were likely contemporaries, on the basis of Babylonian and Hebraic historical accounts. "Hammurabi's Code"[63] contains numerous parallels with the more ancient and comprehensive Torah law.

55. Intercalation is the insertion into a calendar of calculated days or months to align the solar calendar with seasonality. The word *intercalate* is a contraction of the Latin prefix *inter-*, meaning "between" or "among," and the Latin verb *calare*, meaning "to proclaim" or "to call."

56. The Babylonian Talmud in tractate Beraḥot 63A remarks: "Rav Safra said: Rabbi Abbahu would relate: When Ḥanina, son of Rabbi Yehoshua's brother, went to the Diaspora, Babylonia, he would intercalate years and establish months outside the land of Israel. Because Judaism in Israel had declined in the wake of the bar Koḥba revolt, he considered it necessary to cultivate the Jewish community in Babylonia as the center of the Jewish people. Among other things, he intercalated the years and established the months despite the halaḥa restricting those activities to Israel." In the same passage, Ḥanina later says, "Didn't Rabbi Akiva ben Yosef also intercalate years and establish months outside the land of Israel?"
According to Maimonides, in Kiddush HaḤodesh (Sanctification of the Month[s]) (5:1): "All the statements made previously regarding the [requirement to] sanctify Rosh Ḥodesh because of the sighting of the Moon, and [to] establish a leap year to reconcile the calendar, or because of a necessity, apply to the Sanhedrin in the Land of Israel. [For it is they] alone, or a court of judges possessing semiḥah (rabbinical ordination) that hold sessions in the Land of Israel and that was granted authority by the Sanhedrin, [who may authorize these decisions]. [This concept is derived] from the commandment given to Moses and Aaron [Exodus 12:2]: "This month shall be for you the first of months." The Oral Tradition as passed down, teacher to student, from Moses our teacher [throughout the generations, explains that] the verse is interpreted as follows: This testimony is entrusted to you and those [sages] who arise after you and who function in your position. When, however, there is no Sanhedrin in the Land of Israel, we establish the monthly calendar and institute leap years solely according to the fixed calendar that is followed now.
Maimonides (ca. 1135 to 1204 CE), a.k.a. Moses ben Maimom, a.k.a. RAMBAM (a contraction of Rabbi Moses Ben Maimom), was one of the most significant Torah scholars of the Middle Ages. Apart from an encyclopedic knowledge of the Bible and associated writings and commentaries, he was also an accomplished and renowned astronomer, philosopher, and physician. He was the personal physician to the court of Saladin and other royals. His family traced a direct paternal lineage to Judah the Prince, the redactor of the Mishna.

57. According to biblical tradition, centuries before the revelation at Mount Sinai, Abraham, the patriarch of the Jewish people, kept all of the Torah's decrees and ordinances. The Babylonian Talmud in tractate "Yoma," page 28B, reads, "Rav said, and some say Rav Ashi said: Abraham our patriarch fulfilled the entire Torah, even the mitzva of the joining of cooked foods, a rabbinic ordinance instituted later, as it is stated: My Torahs. Since the term is in the plural, it indicates that Abraham kept two Torahs; one, the Written Torah, and one, the Oral Torah. In the course of fulfilling the Oral Torah, he fulfilled all the details and parameters included therein."

58. Genesis 18:10.

59. Shlomo Yitzḥaki, today generally known by the contraction Rashi (RAbbi SHlomo Itzḥaki), was a prodigious and highly acclaimed eleventh-century biblical commentator (1040–1105 CE). His Talmudic commentary is the reference tool of choice for Talmudic scholars the world over. It has spawned more than three hundred "super-commentaries." Rashi does not provide a source for his comment about Passover, but I direct the reader to the appendix for corroborating passages.

60. כעת חיה (ka'es ḥayah) means at this time next year—it was the spring festival of Passover, and on the subsequent Passover, Isaac was born. In Genesis 18:6, Abraham requests of Sarah to make "ugot" for their guests. This very same term describes the Passover matzah in Exodus 12:39: "And they baked unleavened cakes (עוגת מצות) of the dough that they had taken out of Egypt, for it was not leavened, since they had been driven out of Egypt and could not delay; nor had they prepared any provisions for themselves." That Abraham celebrated Passover is an ancient Jewish tradition. Genesis 19:3, just a few passages later, involves Lot (Abraham's nephew) offering his two guests (the same guests/angels that had just visited Abraham and Sarah) ומצות (ooh-matzot), matzah or unleavened bread.

61. "The Calendar Takes Shape in Mesopotamia," Encyclopedia.com, https://www.encyclopedia.com/science/encyclopedias-almanacs-transcripts-and-maps/calendar-takes-shape-mesopotamia.

62. There is considerable debate among scholars as to whether Hammurabi is synonymous with the biblical Amraphel, but that discussion stretches beyond this book's scope.

63. The Code of Hammurabi is a Babylonian code of law originating ca. 1754 BCE. It is one of the oldest and lengthiest deciphered writings, comprising 281 laws. Some of these laws appear on a partially complete stone slab (discovered in 1901 and now located in the Louvre Museum), shaped like a gigantic human finger.

In the *Kuzari*,[64] the author takes to task the subjugating nations of Israel and subsequent civilizations who fail to acknowledge Jewish sources for their acquired wisdom:

> *One cannot fault Greek philosophers for their mistaken beliefs. After all, they are a nation which never inherited Divine wisdom [as it relates to creation] or Torah. The Greeks derive [their wisdom] from the descendants of Yefes,*[65] *who hail from the North. The Divinely inspired wisdom of which we speak,* ***which was passed down from Adam,***[66] *only went to Shem, who was the elite descendant of Noaḥ. This wisdom has never ceased nor will it ever cease from mankind. Even the wisdom which the Greeks do possess was only acquired by them after they conquered the Persian Empire and adopted their wisdom. Persia, in turn, had acquired its knowledge from the Babylonian Empire. Greek philosophers, then, made their appearance only after that point in history when Persia was conquered. And later, when the Roman Empire conquered the Greek Empire, no philosopher of that stature arose ever again.*[67]
>
> *. . . Perhaps you can say the same about Solomon's wisdom [that our knowledge today pales in comparison]. Blessed with Divine inspiration, a masterful intellect, and an inborn disposition, he was able to deal with all areas of knowledge. All the nations of the world—from as far as India—came to him to transcribe his wisdom for themselves.*[68] *Thus, the foundations and fundamentals of all areas of knowledge trace themselves to us [the Jewish people]. They were passed first to the Chaldeans [the Babylonians], then to Persia and Media [northwestern Iran], then to Greece, and then to Rome. With the passage of time and multiple transcriptions, it was forgotten that these areas of wisdom originated with the Jews, and they were instead attributed to the Greeks and Romans.*

On the basis of the aforementioned exegesis and incorporating substantial scholarly research and hypothesis, I postulate (albeit contrary to considerable opposing secular opinion) that it is possible, if not probable, that the Jewish tradition of time measurement was appropriated by Babylonian society and, from there, proliferated to the rest of the ancient world.

For further exploration on ancient Jewish intercalation and its transmission, please refer to the appendix.

THE MODERN CALENDAR

The Gregorian calendar is the world's most common extant solar calendar. Named for Pope Gregory XIII, it has been in continual use since October 1582. It is also the standardized calendar used as a basis for ISO 8601 (International Organization for Standardization[69]—Date and Time Format, which presents the date "universally" as YYYY-MM-DD).

64. One of the most famous works of the medieval Spanish Jewish philosopher and poet Rabbi Judah Halevi, the *Kuzari* was completed around 1140. Originally written in Arabic and subsequently translated into Hebrew, many regard it as one the most important works of Jewish philosophy.
65. One of Noaḥ's three sons.
66. Although no *known* physical or written documentary evidence survives from the epoch of biblical Adam (3760–2830 BCE), the ancient Jewish oral tradition (see earlier footnote about its extraordinary importance) places the origins of the Hebraic lunisolar calendar with the ancestor of humankind. According to this tradition, the knowledge disseminated to Shem (Noaḥ's son) ten generations later. This account appears to stretch credulity, given the 1,500-year separation in the births of Adam and Shem. However, accounting for the enormous life spans of antediluvian (pre-"Flood") man, Jewish tradition has Adam's life span overlapping with Methuselah (the longest-lived biblical figure) and Methuselah's life span overlapping with that of Shem. Shem, the wisest of Noaḥ's three sons and his spiritual heir, established a seminary in the area around Israel. Shem knew Methuselah (and was ninety-eight years old at the time of Methuselah's death, just prior to "the Flood"). Methuselah in turn knew Adam directly. As an aside, Aramaic interpretive translations (Targumim) also identify Shem by the alias Melḥizedek, the king of Salem. Please refer to the appendix for additional discussion on calendric intercalation and transmission.
67. *Kuzari,* First Essay, chapter 63.
68. *Kuzari*, Second Essay (chapter 66); as further evidence of the monumental and astounding wisdom of King Solomon and the degree to which early Jewish society was steeped in the knowledge of mathematics and astronomy, I describe here the earliest known four significant-digit precision for the transcendental number pi (π), being the circumference of a circle divided by its diameter. In the book of Kings (7:23), it reads וַיַּעַשׂ אֶת־הַיָּם מוּצָק עֶשֶׂר בָּאַמָּה מִשְּׂפָתוֹ עַד־שְׂפָתוֹ עָגֹל סָבִיב וְחָמֵשׁ בָּאַמָּה קוֹמָתוֹ וקוה [וְקָו] שְׁלֹשִׁים בָּאַמָּה יָסֹב אֹתוֹ סָבִיב׃. Translated as "And [Solomon] made the basin [sea] of cast metal. [It was] 10 cubits from one lip to the other lip, circular all around; its height was 5 cubits, and a line of 30 cubits could surround it." Notice in the original Hebrew that the word "וקוה" (v'kavah) is to be read "וקו" (v'kav). This is known in Hebrew as Kri UK'siv, when a word is spelled one way yet pronounced another way, and it occurs only a handful of times throughout the Bible. This Kri UK'siv knowledge is ancient, handed down from teacher to student since the time of Moses. There is also an ancient alphanumeric code in Hebrew called Gematria, which assigns a numerical value to a name, word, or phrase on the basis of its letter combination. Every letter in the Torah has a unique numerical value (1 = א ,2 = ב, and so forth). The first letter of the "written"/"read" word is a vav ו, which is a conjunction. We will remove this conjunction from the analysis (as is the norm), leaving the base words קו ("kav"—line) and קוה ("kavah"). The latter (longer) word has a numerical value of 111. The former (shorter pronounced) word has a numerical value of 106. Multiplying 3 (which is 30 cubits / 10 cubits—as found elsewhere in the text) by this ratio (111/106 = 1.047169811320755) gives the product (3 × 1.04 . . .): 3.141509433962264. It is extraordinary that a more than 2,000-year-old text presents pi accurately to four significant digits! Those insufficiently versed in the Hebraic traditions believe that the Jews knew pi only to be 3. Rather, 3 was used by Jewish philosophers as a convenience, since 3 (as an approximation) is over 95 percent the value of pi and sufficed as a proxy for the "real" value of pi.
69. The "International Organization for Standardization," when written as an acronym (e.g., IOS), looks different in different languages. The organization picked ISO as the short form.

September 1752

Su	Mo	Tu	We	Th	Fr	Sa
		1	2	14	15	16
17	18	19	20	21	22	23
24	25	26	27	28	29	30

An unusual sight: Wednesday, September 2, 1752, followed directly by Thursday, September 14, 1752.

The Gregorian calendar replaced its predecessor, the long-lived Julian calendar, and introduced a novel feature to bring the solar calendar into more perfect alignment with the solar year, while also reducing the length of the average year from 365.25 days to a more precise 365.2425 days.

The Julian calendar was itself a modification of the earlier Roman calendar, which comprised just ten months beginning with March. Julius Caesar proposed the modified calendar in 46 BCE, and it became authoritative a year later, lasting more than 1,600 years without significant alteration. The Julian calendar treated solar years as consisting of precisely 365.25 days. Every four years, an intercalated leap day brought the calendar closer in line with celestial reality. In practice, the Julian calendar consists of normal years of 365 days and leap years of 366 days (every fourth year).

However, even with the improved precision of the Julian calendar, an inaccuracy of almost one day per century still arises. By the time Pope Gregory XIII entered the picture, the calendar's steady regression amounted to a gaping imprecision of **fourteen days**. Pope Gregory's desire was to align the calendar with the true vernal equinox (March 21, and not the "exegetical" March 25). As a result, there was no October 5 through October 14 in 1582. Instead, October 15, 1582, immediately followed October 4, 1582—which would make for an excellent trivia question, viz. "What day followed October 4, 1582?"

Indeed, hosts of trivia stem from the Gregorian drift, as I like to call it. Say, for instance, you were born on January 1, 2000. Your day of birth is still Saturday, but the day, month, and solar year under the Julian calendar differ significantly, and your Julian birthday appears as December 19, 1999. What, you say? It should be December 22, on the basis of the ten-day difference between the Julian and Gregorian calendars when they were established (and as noted above). Well, it would be, if not for the addition of the three leap days inserted by the Julian calendar (every century year since 1582) that the Gregorian calendar ignores; however, they both agree on 2000 CE being a leap year (see below). The difference between the calendars increases to four leap days at the turn of the next century, some nearly eight decades hence—therefore, subtracting fourteen days from your current Gregorian birthday in 2100 will yield the correct Julian calendar date.

To reduce future calendar drift, the Gregorian calendar introduced an additional refinement: only century years *exactly divisible by 400* would become leap years. Thus, 2000 CE qualifies as a leap year, but 1900 CE does not.

The British Calendar (New Style) Act of 1750[70] introduced the following clause:

> [1.]. The old supputation [calculation/computation] of the year not to be made use of after Dec. 1751. Year to commence for the future on 1 Jan. The days to be numbered as now until 2d[71] Sept. 1752; and the day following to be accounted 14 Sept. omitting 11 days.

The abovementioned rather cryptic staccato paragraph in eighteenth-century legalese set January 1 as the beginning of the year and stripped September 3–13 (of 1752) from the calendar. Seventeen fifty-one became a short year of 282 days, running from March 25, 1751 (the prior "beginning" of the solar year), to December 31, 1751. This new calendar was to apply to England, Wales, and the colonies henceforward.

70. https://www.legislation.gov.uk/apgb/Geo2/24/23/contents.
71. [sic]; this is how they wrote 2nd.

It took literally centuries for other nations to adopt the now internationally recognized and standardized Gregorian calendar. Greece, as a more recent example, introduced it only in 1923. Japan was not that far off, having introduced it in 1873 (previously utilizing a Chinese-based lunisolar calendar).

MODERN TIMEKEEPING

Concomitant with an increased need for more precise timekeeping, commencing in the fourteenth century the first mechanical clocks began appearing, and the adoption of fixed-length hours became more widespread. Three centuries later, in 1656, the pendulum[72] clock made its successful debut.

By the seventeenth century, pocket watches had evolved quite significantly since the fifteenth-century invention and introduction of the mainspring. However, the pendulum clock still reigned supreme in terms of accuracy until the 1930s, with the advent of the quartz oscillator. Later chapters differentiate the types and parts of watch movements in detail.

Invented after World War II, the atomic clock[73] has continued timekeeping improvement to astonishing levels of precision. In 2015, a strontium clock developed at the National Institute of Standards and Technology in collaboration with the University of Colorado Boulder made headlines when scientists revealed that it was accurate to within one second in *fifteen billion years*.

THE SMALLEST UNIT OF TIME

We are quite familiar with the second. There are *twenty-four* × *sixty* × *sixty* seconds in every day. That provides 86,400 opportunities to smile or take a deep breath or laugh out loud. We even incorporate the word into our speech as a time unit of variable length, roughly meaning a moment: "Just a second, please."

We all are also probably aware of the millisecond (one thousandth of a second), microsecond (one millionth of a second), and even the nanosecond (one billionth of a second). Telecommunication and computer equipment often operates at frequencies where these tiny fractions of a second are meaningful.

A nanosecond is 1×10^{-9} seconds. Written out, it looks like this:

0.000000001 seconds.

What is the smallest *physically meaningful measure of time*? It turns out it is *much* smaller than a nanosecond. It is a unit of measure called "Planck time"; the time that light takes to travel one Planck length.[74] We know this is the smallest possible unit of time, because smaller time units have no use in physics (as we currently understand the field of study). A Planck time unit is approximately $5.391247(60) \times 10^{-44}$ seconds. It is hard to wrap one's head around a number this infinitesimal. Here it is in standard decimal notation, for you to see in all its tiny majesty:

0.00000000000000000000000000000000000000000005391247 seconds.[75]

Between the nanosecond and the "Planck time unit," a whole host of fancy-named units of time reside. There is the "jiffy": the amount of time light takes to travel one fermi[76] (about the size of a nucleon) in a vacuum[77]—that is, 3×10^{-24} seconds. Coming in at more than three hundred times longer duration is the zeptosecond (10^{-21} seconds); 850 zeptoseconds is the smallest currently measurable fragment of time using a strontium atomic clock.

Next time you use the term "moment," consider its proper archaic use: an unexpectedly quantifiable measure of time equivalent to one-fortieth of a solar hour—ninety seconds on average— and used by medieval astronomers to compute astronomical motion.

72. The pendulum is a weight connected to a pivot that oscillates freely at a precise interval dependent on the length of the rod to which the weight connects. The invention of the pendulum as a timekeeping device attributes to Christiaan Huygens, who patented the design in 1657.

73. Atomic clocks measure the electromagnetic signal (frequency) that electrons emit when they change energy levels within atoms. In 1968, the International System of Units (SI) defined the second as the duration of 9,192,631,770 radiation cycles, concomitant with electrons transitioning between two energy levels of the ground state of caesium-133.

74. In physics, the Planck length, denoted ℓP, is a unit of length that is the distance light (in a perfect vacuum) travels in one unit of Planck time. It is equal to $1.616255(18) \times 10^{-35}$ m. It is the fundamental natural unit for length derived from dimensional analysis of the universal gravitational constant, the speed of light, and Planck's constant divided by 2 pi. Each of these constants has some variability in precision; pi is always approximated because it is irrational. Renowned physicist Max Planck derived the Planck length in 1899. It is believed to be the scale at which quantum gravitational effects become apparent.

75. The number 60 in parentheses (60) (a.k.a. shorthand error notation) regarding the Planck time unit, elaborated on a paragraph prior, indicates an uncertainty in the measured value. The number 60 (being a two-digit number) applies to the (two) least significant digits, here being $5.391247^{-44} \pm 0.000060^{-44}$ seconds.

76. 10^{-15} meters.

77. There are other definitions for the "jiffy."

On the larger end of the time spectrum (so to speak), there is the gargantuan yottasecond, or 10^{24} seconds, which is about 31.7 quadrillion years. That's still well below a googol *year* (10^{100}), which is when proton decay and heat death is conjectured to occur as the ultimate fate of the present universe.

In 2020, physicists at Pennsylvania State University[78] modeled a universal clock that quantized (subdivided into measurable increments) time itself. Just as the Higgs field gives rise to the Higgs boson, so too would this universal time field potentially generate a chronon.[79] The model suggests that atomic clocks would sometimes desynchronize with the universal clock, eventually leading to two atomic clocks potentially disagreeing about the measurement of a span of time. This all leads to the determination of an upper bound for the fundamental period of time at 10^{-33} seconds. It could be smaller than this, but as of mid-2020, the smallest unit of time must be at least this minuscule.

A perpetual-motion machine? Almost. Invented in 1928, the Jaeger-LeCoultre Atmos (ca. 2014, MSRP: $8,600, ref. 511.72.01) runs on air—literally. The case is hermetically sealed, and even the smallest variation in ambient temperature is sufficient to energize the power reserve. A single degree (centigrade) rise or fall in ambient temperature will provide sufficient energy for forty-eight hours of use.

Inside the clock is an ethylene-chloride gas capsule. With temperature variations, this gas expands or contracts, making the capsule expand and contract too. A short chain connected to the wall of the capsule winds the mainspring through a reciprocating motion. Powered by calibre 562 with 231 lubricant-free components, its massive balance (suspended from a superfine Elinvar wire) vibrates at a leisurely 120 vibrations per hour. Its moon phase is accurate to within one day in 3,821 years.

78. Garrett Wendel, Luis Martínez, and Martin Bojowald, "Physical Implications of a Fundamental Period of Time," *Physical Review Letters* 124, no. 24 (June 19, 2020): 241301.

79. A chronon is a proposed quantum of time that is a discrete and indivisible "unit" of time as part of a hypothesis that proposes that time is not continuous.

Décembre
Novembre
Octobre
Janvier
Février
Août
Juillet
ATMOS
SWISS MADE
JAEGER-LECOULTRE
ATMOS

THE PHILOSOPHY AND SCIENCE OF TIME (AND SPACE)

"But man postpones or remembers; he does not live in the present, but with reverted eye laments the past, or, heedless of the riches that surround him, stands on tiptoe to foresee the future. He cannot be happy and strong until he too lives with nature in the present, above time."

–Ralph Waldo Emerson, *Self-Reliance*

What is time? It seems like a simple-enough question until one really tries to meditate on it. Add in Albert Einstein's theories of relativity, and now space and time unify as space-time—"pliable" interlinked concepts. Well before Einstein, philosophers[80] tackled the nature of time.

Do space and time exist independently of mind? Why does time (apparently) move in only one direction? Do instances other than the present moment exist? Was there a moment without an earlier one? Are there points in time with zero duration? What neural (brain/mind-related) mechanisms are involved in our experience of time? It is difficult to argue that we experience *something* analogous to the passing of time. But what exactly is it that we are experiencing?

This contentious topic has concerned scientists and philosophers since antiquity, and while I attempt to provide an overview here, the subject is extensive enough to warrant its own book. Indeed, many excellent references are available.[81]

Before embarking on philosophical analysis, a few words of caution. The following section is complex. I have made my best effort to explain varying philosophies without losing nuance, while still (I hope) making the text comprehensible. Philosophical interpretation is convoluted; it often leaves one struggling with more questions than answers. The philosopher/metaphysician is likened to a "blind man in a dark room looking for a black cat (or hat) that is not there." The quote's absolute origin remains unknown, but I think the sentiment is rather apt.

80. The word "philosophy," like the word "horology," is Greek in origin. Philosophy is the study of the fundamental nature of knowledge, reality, and existence. Its Greek roots are *philo* (love) and *sophia* (wisdom)—that is, love of wisdom.

81. There is, for instance, the entertaining if somewhat controversial *The End of Time: The Next Revolution in Our Understanding of the Universe*, a popular science book published in 1999. In the book, author Julian Barbour argues that time exists merely as an illusion, and our insistence on it in physical theory has held science back from the scientific revolution that awaits. For Barbour, time is nothing but change.

TIME AND THE CONSCIOUS EXPERIENCE

Whatever our consciousness may be—and that is a subject of intense debate[82]—it evidently exists within time. When the sun rises at 6:40 a.m. (on September 20, 2020, in New York City—as displayed by the hands on my wristwatch), the visual experience of our bright waxing home star *corresponds* to the time of sunrise. The appearance of the outside world and the reactions within our inner consciousness seemingly correspond within the same temporal[83] dimension. But how exactly does our consciousness work? How is it that these experiences are even possible?

We recall temporal intervals and memories of past states of consciousness quite easily. Moreover, while consciousness exists within time, time may exist within consciousness. When we listen to the opening bars of Beethoven's fifth symphony—"da da da daaa . . . da da da daaaaaaa"—we are able to discern each fleeting note as it generates our experience. Surely then, we can conclude that our awareness encompasses the temporal interval (roughly seven seconds[84]) of that experience—that is, if we hear da "da da daaa . . . ," the entire experience must have entered our awareness.

Not. So. Fast.

This is where all the mind-bending trouble begins.

What is the present? What is *the now*? Try to hold on to it. You cannot. It is fleeting. It is ephemeral; the moment we grasp it, it has already slipped away into an ever-growing "past." Our present awareness is only of "this" particular "slice of instance." Now. Now too. Our awareness apparently lacks persistence and temporal depth.

Philosophers describe three primary models[85] to express the aforementioned temporal awareness paradox—that is, experiencing an event yet realizing that the present may be a mere infinitesimal slice tending toward zero duration.

82. Numerous definitions exist for consciousness. Most definitions would define it as "sentience or awareness of one's surroundings." "Your consciousness" (whatever that is) is aware right now, one hopes, that "you" are reading this book. The nature of consciousness, which is everything you experience, may be even more enigmatic than the nature of time. Most scholars accept consciousness as a given and attempt to understand how it relates to the objective world of science.

83. Adjective: of or relating to time.

84. Whether it is Sir John Eliot Gardiner, Carlos Kleiber, or Leonard Bernstein conducting the five opening bars of (arguably) the most recognizable musical piece of all time, Beethoven's sublime Symphony No. 5, the duration is somewhere between six and eight seconds, depending on the conductor's interpretation of "allegro con brio" or "fast tempo, with spirit."

85. Barry Dainton, "Temporal Consciousness," in *The Stanford Encyclopedia of Philosophy* (Winter 2018 Edition), ed. Edward N. Zalta, https://plato.stanford.edu/archives/win2018/entries/consciousness-temporal/.

In the **extensional model**, as the name obligingly implies, our experience consists of extended interwoven clusters. Our immediate experience of change and succession occurs within "specious present[86] moments which are themselves extended through time."[87] Using Beethoven's masterpiece as our "experience" example, the "extended moment" encompasses all five bars.

The **retentional model** combines retentions of the recent past with immediate experience. Our apparent immediate experience of change and persistence "is in fact packaged into momentary (or extremely brief) slices of experience."[88] "In other words, the sense of temporal continuity is provided by adding the intentional presence of the earlier and later phases of the object, to the primal impression (the now)."[89]

Last, the **cinematic model** parallels the motion-picture experience of static images appearing in rapid succession. In our Beethoven example, each note, or even each part of a note, represents a "slice of time." Augustine,[90] who subscribed to presentism—acknowledging only the present as real—would probably have embraced this position. Contemporary adherents of the cinematic model, however, must explain how we possess awareness of successive experiences if our consciousness consists of nothing more than the perception of each infinitesimal instant. How does our consciousness weave together an "experience," if all we perceive are tiny snapshots of the world?

Immanuel Kant[91] and fellow philosophers in the late eighteenth and early nineteenth centuries, as supporters of the retentional model, require cognitive thought as a precondition to perceptual experience. Again referencing the Beethoven snippet, our cognitive thought must somehow "hold in place" the experience of the first few bars before "experiencing" the next bars. At each instant, one is aware of a momentary sound, yet simultaneously, one is aware of the preceding series of sounds. When we hear "da da da daaa . . . ," we may only "now" be aware of the last note, but the three preceding notes persist in cognition, allowing us to make sense of the entire musical phrase.

The extensional model requires temporal distribution of successive experiences (one experience follows the next, spread out over time). Because experience extends in time, it presents events that can unfold in time. The *now* (specious present) embodies a brief duration and can therefore interweave nonsimultaneous events. However, advocates of this approach have to explain how successive earlier experiences together with the present are simultaneously experienced. Currently, you are experiencing only *now* and a recollection of the recent now. How can you experience both the now and this "remembrance" simultaneously, since you are currently aware only of the present's *now*?

Theologians may otherwise be inclined to accept formerly *lived experience* as an extensional model, but the doctrine of perpetual creation[92] implies a cinematic model. Unfortunately, as mentioned, the cinematic model is open to its own philosophical opposition, not least in finding an explanation for adjoining experiences that are otherwise distinct. How can change and succession feature in immediate experience?

Bruno Mölder does not think that current philosophical modeling provides adequate elucidations of the underlying cause of time consciousness, instead relying on explaining only possible temporal mechanisms:

> *I conclude that* [*these models of time consciousness*], *although they evidently aim to provide explanations, none of the*

86. "The term 'specious present' was first introduced by the psychologist E. R. Clay, but the best-known characterization of it was due to William James, widely regarded as one of the founders of modern psychology. He lived from 1842 to 1910, and he was professor both of psychology and of philosophy at Harvard. His definition of the specious present goes as follows: 'The prototype of all conceived times is the specious present, the short duration of which we are immediately and incessantly sensible' (James 1890). How long is this specious present? Elsewhere in the same work, James asserts, 'We are constantly aware of a certain duration—the specious present—varying from a few seconds to probably not more than a minute, and this duration (with its content perceived as having one part earlier and another part later) is the original intuition of time.' This surprising variation in the length of the specious present makes one suspect that more than one definition is hidden in James' rather vague characterization." Robin Le Poidevin, "The Experience and Perception of Time," in *The Stanford Encyclopedia of Philosophy* (Summer 2019 Edition), ed. Edward N. Zalta.

87. Barry Dainton, "Temporal Consciousness," in *The Stanford Encyclopedia of Philosophy* (Winter 2018 Edition), ed. Edward N. Zalta, https://plato.stanford.edu/archives/sum2019/entries/time-experience/.

88. Ibid.

89. Bruno Mölder, University of Tartu, Ülikooli 18, 50090 Tartu, Estonia, *Procedia—Social and Behavioral Sciences* 126 (2014): 48–57. Parenthesis added.

90. Augustine of Hippo (354–430 CE), a.k.a. Saint Augustine, was a theologian and philosopher in Roman North Africa. His writings influenced Western philosophy and Christianity.

91. Immanuel Kant (1724–1804) is one of the central figures in modern Western philosophy. Kant was neither an empiricist nor a rationalist. Empiricists and rationalists disagreed on whether all concepts derive from experience and whether humans can have any substantive prior knowledge of the physical or metaphysical worlds. Kant synthesized the insights of both movements in his new critical philosophy. By doing so, he led a new era of German idealism and late modern philosophy. Kant continues to exercise significant influence in metaphysics, epistemology, ethics, political philosophy, and aesthetics.

92. Numerous biblical commentators support the position that G-d renews Creation perpetually and that were He not to do so, the entire universe would instantly cease to exist.

common models of explanation are readily applicable to them. However, they could be regarded as mechanistic explanations in a loose or extended sense. Namely, they purport to describe the mechanism of time consciousness but do so without stepping out from the phenomenological[93] *domain.* ***They aim to answer some how-possibly questions about our experience but they are still far from giving how-actually explanations****, which "describe real components, activities, and organizational features of the mechanism that in fact produces the phenomenon."*[94] *We should perhaps only reckon that building a philosophical model at the level of experience does not give us the whole mechanism of how we experience change and other temporal properties. It is likely that the real causal mechanism is to be understood with the help of empirical psychology and neuroscience and the role of philosophy is more modest than one might want to assume.*

TIME AND CHANGE

George Santayana (1863–1952), a Spanish American philosopher, novelist, and poet, famously wrote, "The essence of nowness runs like fire along the fuse of time, but the particular spark is different at each point. The various contents of these various *nows* therefore combine perfectly to form the unchangeable truth of history."[95] That sentiment embellishes upon Plato's much earlier "Time is the moving image of reality."[96]

The realists[97] posit that time and space exist apart from the human mind. This outlook concurs with the Platonic view of time: time is but a reflection of the rotation of the celestial bodies.

Idealists hold the opposite opinion: objects cannot exist independently of mind. This is the reductionist[98] view of time supported by Aristotle[99] and (later) Gottfried Wilhelm Leibniz:[100] time embeds in motion and is meaningful only with respect to events within its flow. The underlying basis of the realist/idealist argument is whether time can exist independently of change.[101]

Philosophical substantivalists would surmise that time flows equally and without regard to anything external. Leibniz and other relationists would say that time is merely the relations between events. The dispute remains unresolved.

Raymond Tallis[102] encapsulates the modern philosophical impasse:

> *The relationship between time and change remains elusive. Giving time priority over change, and imagining time continuing in the absence of events, have unsatisfactory consequences. It is no less unsatisfactory to see time as generated from, or subsisting in, the relations between events, if only because this leaves us with the seeming impossibility of characterizing the nature of that relationship without mentioning the word "time."*

Theologians from antiquity to present day take the view that time exists only within the created universe, and that G-d operates outside time.

Many modern philosophers, as previously noted, contend that time itself is an illusion.

In explaining his argument for the *unreality of time*, Scottish idealist philosopher John McTaggart[103] introduced "A-Series" and "B-Series" as two descriptions for the temporal ordering of events. The A-Series locates each event relative to the present; the B-Series is created with no attention paid to the present, but only to what occurs before what.[104]

The A-Series orders events as past, present, and future, with only future and past allowing for degrees of magnitude (i.e., more "pastness" or more "futureness")—all existing within a series of temporal positions in continual transformation.

93. Relating to the science of *phenomena* (a fact or situation that is observed to exist or occur), as distinct from that of the *nature of being*.
94. Carl F. Craver, *Explaining the Brain: Mechanisms and the Mosaic Unity of Neuroscience* (Oxford: Oxford University Press, 2007).
95. *Realms of Being* (1932, p. 491)
96. Timaeus (ca. 360 BCE) is one of Plato's dialogues in the form of a monologue, attempting to describe how the world came into being.
97. A philosophical grouping of like-minded thinkers who believe things exist independently of experience.
98. Time can be reduced to a discussion about temporal relations among objects and occurrences.
99. Aristotle (ca. 384–ca. 322 BCE), together with his teacher Plato, is considered the "Father of Western Philosophy." His writings cover a multitude of topics, including biology, economics, esthetics, ethics, linguistics, logic, metaphysics, music, physics, poetry, politics, psychology, rhetoric, theater, and zoology.
100. Gottfried Wilhelm von Leibniz (ca. 1646–ca. 1716) was a prominent German logician, mathematician, and natural philosopher. Leibniz's greatest accomplishment was conceiving the ideas behind differential and integral calculus, independently of Isaac Newton's concurrent efforts.
101. Richard Feynman (1918–1988), noted American theoretical physicist, is purported to have quipped, "Time is what happens when nothing else does."
102. Prof. Raymond Tallis, "Time & Change," *Philosophy Now* 115 (August/September 2016).
103. Ca. 1908. McTaggart is probably the most notable supporter of this approach, but there are others.
104. Bradley Dowden, "Time," in *Internet Encyclopedia of Philosophy*, https://iep.utm.edu/time/.

[More Distant Past]
[Distant Past]
Past
Present
Future
[Distant Future]
[More Distant Future]

The above series of temporal events conjectures that an event is first part of the future, then part of the present, and then part of the past. These events map to something outside time. B-Series, on the other hand, predicates that an event that is earlier than another event will always be earlier than another event. Events are not described as past, present, or future, but rather as "earlier than" or "later than."

Earlier than : Later than

The difference between A- and B-Series is more discernable when we look at these two perspectives by means of the principle of temporal parity. This notion holds that contrary to appearances, all "times" really exist in parity (on the same "dimensional plane"). No one time has a better claim to "the present" (whatever that concept means) than another does. This accords with B-Theory (represented by the B-Series of events) but contradicts A-Theory (represented by the A-Series of events).

Those who support the B-Theory of time argue that time is without tense. These "eternalists" postulate that all existence in time is equally real and there is no objective flow of time. We humans just happen to perceive a world where time passes, but this experience is illusory.

A-Theory supporters align with an ordered tensed universe—an increasingly minority view. The *growing-block* universe theory of time embodies some properties of A-Theory—as time passes, more of the world comes into existence. The block universe grows as the present time slice adds to the past block. A-Theory supporters insist that the passage of time is real and not mind dependent.

Richard M. Gale, in his paper "McTaggart's Analysis of Time,"[105] describes why McTaggart concludes that time is contradictory and must therefore not exist:

> Every event in the A-Series, assuming that there is no first or last event, has the mutually incompatible properties of being past, present, and future,[106] which is a contradiction. The obvious reply to this seeming contradiction is to say that no event has two or more of these incompatible properties at the same time, but rather has them successively at different moments of time. But this reply will not do, since it involves either a vicious circle or vicious infinite regress.

The vicious circle of which he speaks *supposedly* arises from the introduction of a second-order time series necessary to explain away the contradiction of an event in the first-order time series. Now the second-order time series requires explanation, necessitating a third-order time series, and so on and so forth as the process extends ad infinitum. In summary, a new A-Series is necessary to explain the temporal contradictions in a previous A-Series, which is quite obviously circular.

Earlier, McTaggart has written:[107]

> Positions in time, as time appears to us prima facie, are distinguished in two ways. Each position is Earlier than some and Later than some of the other positions. . . . In the second place, each position is either Past, Present, or Future. The distinctions of the former class are permanent, while those of the latter are not.

An abstracted representation of "Animal Locomotion," plate 626, circa 1887, by Eadweard Muybridge. Each "frame" illustrates how more of the universe "comes into being" as the overall "block" grows.[108]

105. *American Philosophical Quarterly* 3, no. 2 (April 1966).
106. In other words, an event like tying one's shoes could be past, present, or future depending on one's temporal perspective.
107. J. M. E. McTaggart, *The Nature of Existence*, vol. 2 (Cambridge, UK: Cambridge University Press, 1927), 9–10.
108. Californian politician and railroad tycoon Leland Stanford—yes, *that* Stanford (cofounder of Stanford University)—had a passion for racehorses. He pondered whether a horse was ever completely aloft or if one or more limbs always made ground contact. Eccentric photographer Eadweard Muybridge, who years prior thought the task unverifiable on the basis of prevailing technology, eventually settled the question: all four limbs are aloft simultaneously. It was June 15, 1878, in Palo Alto. On the basis of accounts of previous attempts, reporters and spectators gathered. Muybridge lined up twelve state-of-the-art cameras. By linking twelve individual trip wires (in the horse's path) to twelve distinct cameras, Muybridge was able to capture twelve still images in under half a second. The image shown here displays the first six still frames (of twelve) from one of Muybridge's later locomotion studies. Muybridge may have not known it at the time, but he set the stage for the advent of the motion picture. Contemporary animators and artists such as Degas and Eakins referenced Muybridge's images for proper equine anatomical posture while in motion.

Gale rephrases McTaggart's distinction.

> First, we can say that one event is earlier (or later) than some other event; and, second, we can say that some event is now past (or present, or future).

He explains that these are fundamentally different ways to make a temporal determination. The first part of the sentence—"*First, we can say that one event is earlier (or later) than some other event*"—will always have the same properties of truth *irrespective* of when uttered. A simple example: "I get out of bed" is always earlier than "I put on my shoes" (unless you wear your shoes in bed). Given any two nonsimultaneous events, x and y, either x is earlier (or later) than y or y is earlier (or later) than x.

We infer that the only change possible is a change in an event's position in the A-Series. McTaggart concludes that since the B-Series cannot give us change, although it (the B-Series) is necessary, it is not sufficient to describe the reality of time. Without a temporal A-Series, there can be no B-Series. With neither, there can be no time. Time does not exist. Gale argues with McTaggart's assertions and summarizes:

> If we take the A-Series to be determined by an unqualified past, present, and future, i.e., a past, present, and future that do not admit of degrees, then it can be shown that the B-Series cannot be reduced to the A-Series, this being due to the fact that the B-relation "earlier than" cannot be defined in terms of an unqualified "past, present, and future."

The second part of the sentence—"*and second, we can say that some event is now past (or present, or future)*"—may prompt statements with different truths *depending* on when uttered. Another simple example: "I get out of bed earlier than I put on my shoes" is equivalent to saying:

"Getting out of bed is past and putting on my shoes is present," or
"Getting out of bed is past and putting on my shoes is future," or
"Getting out of bed is present and putting on my shoes is future," or
"Getting out of bed is more past than putting on my shoes," or
"Putting on my shoes is more future than getting out of bed."

Gale expanded McTaggart's concept of the A-Series by adding the latter two statements, which affix qualifiers ("more") to the degree of "pastness" or "futureness" of an event. By doing so, he aligns an "impure" version of the A-Series with the B-Series. Gale contends that trying to align a pure A-Series with the B-Series is futile.

David Sanford[109] tries to demonstrate the invalidity of McTaggart's contradictory A-Series argument and specifically attempts to cast doubt on Gale's analysis by removing "tensed verbs." He posits that McTaggart "can be answered without espousing any philosophical thesis about the nature of time." He explains that the "conclusion that the A-Series is self-contradictory does not follow from McTaggart's premises." Sanford reframes these "contradictions" as vacuous truths. "To say that an event *will be* past and *has been* future is to say nothing. Only one moment *is* present and any moment present when this moment *is* present, is this moment. . . . True, the moment which *is* present is also past and future in that it *will be* past and *has been* future. But there is no inconsistency here."

The aforementioned current metaphysical dispute is not new or unique; in fact, it has been debated in one form or another since antiquity. Ancient Greek philosopher Parmenides believed that reality is timeless and unchanging (akin to B-Theory). Heraclitus, his contemporary, thought that the world is transient and in a constant state of fluidity (akin to A-Theory).

THE SPEED AND DIRECTION OF TIME?

Historically, time measurement correlated to direct observations of the heavenly spheres; the rising of Sirius, the eternal procession of night following day and day following night, the position of the sun in the sky, the phase of the moon, the marking of the meridiem, or the shadow cast by the sundial's gnomon.

Present-day measurements utilize complex instruments accurate to billionths or trillionths of a second, but to scientists and philosophers, the physical nature of time remains intractably enigmatic. We remember the past

109. David H. Sanford, "McTaggart on Time," *Philosophy* 43, no. 166 (1968), 371–78.

but not the future. Omelets do not become eggs. Shards of broken glass do not suddenly rearrange into a pristine champagne flute. The arrow of time appears to flow in only one direction as entropy[110] increases; time passes and *disorder* increases. A fundamental question then arises: Why or how was the universe so immaculately wound up? If disorder only increases (if higher entropy states are overwhelmingly more statistically likely—see footnote), how do we end up with such a perfectly calibrated functioning universe? Why do we ever have any order at all? If order is unlikely, where does it come from? The extant universe should instead present itself as a cosmic soup of cool randomized particles. In summary, if we trace the entropy puzzle back to the "start," we can hypothesize (scientists call this the "past hypothesis") that the early universe had extraordinarily low entropy (high order). While theology advances an Omnipotent, Omniscient, and Omnipresent Creator in explanation, science does not yet offer any concrete answers.

Some modern cosmologists suggest that the observable universe is contained within a much bigger multiverse and that the Big Bang was not the beginning. Moreover, physics does not require a forward-moving arrow of time for its equations to make sense.

How time feels to the observer is always the same, but since Einstein's relativity revelations, we know that the space-time path traced by the observer affects the measurement of elapsed time. Time's arrow always follows the same direction, but not always at the same "speed."

A *special relativity* thought experiment called the Twin Paradox will help elaborate. Consider identical twin brothers. Twin A gets into a rocket ship traveling at near light speed and ventures into the depths of space. The other brother, Twin B, remains on earth.

When Twin A, who embarked on his high-velocity space adventure, returns, he discovers that his brother, Twin B, has aged substantially, yet he, Twin A, has not. Einstein and Max Born cite gravitational time dilation resulting from the effects of acceleration to explain this effect. Einstein considered time dilation to be a natural consequence of special relativity and not a paradox at all.[111]

110. Entropy: loosely defined as the increase in disorder, chaos, or decay. Ludwig Boltzmann (Austrian physicist and philosopher, 1844–1906) provides a more specific definition (simplified here to ease understanding): entropy is a measure of the number of possible arrangements in a system. As explained by engineer Arvin Ash, physicists understand the three apparently separate concepts—entropy, information, and time—as being interrelated. Almost all of the laws of physics work both forward and backward (they are time reversible), except entropy (as it relates to the second "law" of thermodynamics), which seemingly works only in the forward time direction. I write "seemingly" because there is an infinitesimally tiny (statistically impossible) chance ($10^{150,000,000,000,000,000,000,000}$ or 10 to the power of 150 billion trillion) that the system could revert to its initial state (or any other state, for that matter). Sir Arthur Eddington (astronomer, physicist, and mathematician, 1882–1944) called entropy "time's arrow." Entropy's increase, he speculated, is responsible for time's flow. Since disorder in a system is (overwhelmingly) likely to increase, time is most likely to flow forward. But on the basis of the aforementioned description, one could logically conclude that there is a very tiny possibility that time could "proceed" backward if entropy would reverse itself. However, we do not see this in "reality." According to Ash, "It is not clear that entropy is responsible for the one-way flow of time. Because if it was, then systems where entropy was decreasing, like the inside of a refrigerator, should experience a backward flow of time. But this doesn't happen. Time flows forward in our universe presumably everywhere. I don't think time can be reversed in a universe like ours where overall entropy is always increasing. However, the question remains whether time would flow backward in a universe where entropy of the entire universe was decreasing. This is possible—and could provide an opening for a science fiction story."

Entropy was at its all-time lowest state around the Big Bang's commencement. Time's forward flow and causality, therefore, *in this framework at least*, is a consequence of entropy's increase resulting from the second "law" (read "law" as overwhelming statistical tendency) of thermodynamics.

111. Although it's unnecessary, I have depicted the paradox in an accelerative framework to ease understanding. Schiffmiller (see acknowledgments) favors a twin-paradox scenario without accelerations.

The definition of "special" in special relativity (SR) is that the acceleration is zero. Einstein came up with SR in 1905. It took him until 1917 or so to come up with equations for the relativistic effects in the presence of an acceleration. That is the "general theory of relativity." The books I have on SR present the twin paradox without accelerations. One twin is on Earth, and the other whizzes by Earth in a spaceship moving at a constant velocity. As it passes, they exchange a communication signal that registers and synchronizes their clocks. The spaceship continues, and a time later, another signal is sent and they learn their clocks are no longer synchronized. The "paradox" is that when one body moves with respect to another without acceleration, there is no physical difference and no way to tell if A is moving and B is not, or the reverse. Therefore, why should one age versus the other? This is called reciprocity. The explanation of the oddity is that one twin experienced several accelerations, so their experiences were not symmetric. This is similar to two people a distance apart. A looks "small" to B, while B looks "small" to A. The "triplet" paradox is often presented to strengthen the paradox.

TWIN PARADOX

3 In our hypothetical scenario, the spacecraft successfully accelerates to 99.999% the speed of light *(contracting in length as it does so)* and slowing down time for Twin A by a factor of **224**. For every one hour Twin A experiences in the spacecraft, Twin B experiences 224 hours (9 days, 8 hours) on Earth.

2 With Twin A onboard, the spacecraft leaves Earth headed for space at a relatively pedestrian 17,600 miles per hour. Twin B remains on Earth. The spacecraft speeds up rapidly.

4 After nearly three months (from Twin A's perspective), the spacecraft returns to Earth, slowing down from near lightspeed to land safely.

Twin B has aged 50 years, while the space-faring **Twin A has aged a mere 3 months**. The 75-year-old twin welcomes home his (still) 25-year-old "time-traveling" sibling with a hug.

1 Twins, 25 years old, give each other one last hug before Twin A (with the red sweater) embarks for the future.

Time dilation has been verified experimentally countless times by using carefully calibrated atomic clocks on satellites and aircraft.[112] On a very practical level, GPS satellites require daily nanosecond calibration to eliminate the effects of time dilation caused by their relative velocities. Were it not for this daily correction, your GPS device could not accurately determine your location and you would find that you were many miles off course.[113]

So, did time have a beginning? Theologians contend that it began at the very moment when G-d created The Universe. G-d Himself is timeless and completely independent of a time-bound Universe. As the Torah relates, "In the beginning, G-d created the Heavens and the Earth." Naḥmanides[114] expounded this opening phrase, declaring: "Instead He brought forth from total and absolute nothing[115] a very thin substance devoid of corporeality but having a power of potency, fit to assume form and to proceed from potentiality into reality. This was the primary matter created by G-d." Remarkably, Ramban wrote his commentary almost one millennium ago and Kabbalistic[116] tradition goes back even further than that. Remember, this was long before the discovery of what modern cosmologists now call the "Big Bang," an expansionary epoch radiating from a singularity at the commencement of the current universe.

I encourage you, dear reader, to explore the metaphysics of this topic further.

112. What really bakes one's noodle is considering time from the perspective of a massless particle, such as the photon. From its perspective, traveling from one "end" of the observable universe to another (a distance of many tens of billions of light-years), *no time passes whatsoever*. For the massless particle, everything is instantaneous.

113. The decreasing effects of gravity as altitude increases mean that time passes more slowly closer to sea level, and although the effects are minuscule, they are measurable and can have deleterious effects on sensitive timing equipment. W. Wayt Gibbs (*Scientific American*, September 2002) writes: "Clocks at the top of Mount Everest pull ahead of those at sea level by about 30 microseconds a year. Raising a clock 10 centimeters will change its rate by one part in 10^{17}. And elevation is relatively easy to measure, compared with variations in gravity caused by local geology, the tides, or even magma shifting miles underground."

114. Moses ben Naḥman (1194–1270) was a leading Jewish scholar, philosopher, physician, kabbalist, and biblical commentator. His commentary on the Torah (five books of Moses) is his final and best-known work.

115. אבל הוציא מן האפס הגמור המוחלט יסוד דק מאד, אין בו ממש, אבל הוא כוח ממציא, מוכן לקבל הצורה ולצאת מן הכוח אל הפועל. והוא החומר הראשון,
(English translation—in body text—by Charles B. Chavel, *Ramban (Nachmanides): Commentary on the Torah* (5 Vol. Set), Vol. 1, p. 23)
Regarding בְּרֵאשִׁית, (the literal first word in the Torah), [t]he meaning of the letter ב at the beginning of the word בְּרֵאשִׁית is to indicate a time, just as in the word ביום, "on a certain day." The reason for this is that time itself was created. Time was the very first concept that was created before anything of substance, anything material. . . . The author proceeds a few paragraphs later to describe רֵאשִׁית (the first word of the Torah minus its opening letter). He explains that the Torah should have begun with the word for "first" which is ראשון. By using רֵאשִׁית, it describes "an absolute beginning, a beginning which had not been preceded by anything else at all of whatever category."
(English translation of Haketav Vehakabbalah 1,1, page 1, *Torah Commentary* by Rabbi Yaakov Tzevi Mecklenburg; translation by Eliyahu Munk, 2001)

116. The literal translation of the word Kabbalah is "that which is received." Kabbalah has three "branches": the theoretical—primarily concerned with describing inner reality; the spiritual—the different worlds, dimensions, and souls; and the meditative—use of the Divine Names and permutations to influence natural events. Most of the latter (spiritual) category remains completely unpublished in any language and only verified (and hidden) mystics know these sacred truths. It is also **dangerous to pursue by the "uninitiated and unqualified,"** and one is **strictly forbidden from studying it without prior authorized rabbinic permission.**
The main work of "Theoretical" Kabbalah is covered by such sacred texts as the Zohar and other mystical works of similar ilk. These works too (even in English) should only be studied by someone authorized and sufficiently learned to do so.
More than this cursory introduction to Kabbalah extends beyond this book. Although the study of Kabbalah may prove fascinating and enlightening, **the reader is once again severely cautioned against reading or interpreting Kabbalah in Aramaic, Hebrew, or English** (or any language for that matter) **without the specific permission, guidance, and assistance of a generally accepted and competent Orthodox Halaḥic authority.**

WHAT WE TALK ABOUT WHEN WE TALK ABOUT TIME[117]

> "Time is what we want most, but what we use worst."
>
> –William Penn

Late into my authoring of this book, my son inquired, "Dad, why don't you talk about the South African expression 'just now'? Given the other time-related esoterica in the book, it would fit right in." Of course, he's right. "Just now" is one of the most perfectly expressed antithetical terms in the English linguistic register. When asked, "When will you get to that task?" a common South African response is "just now." Don't be fooled; it does not mean at this moment, as the dictionary definition would contend.

When I first visited the United States, I found that my American colleagues were stupefied that South Africans saying "just now" could mean a few seconds, a few minutes, half an hour, or even more; it is a relative time span and usually dependent on both the specific circumstances and the personalities involved. It is a delightful expression but is seldom used fittingly by nonnative South Africans. Of course, to an American, it sounds like gibberish; "now" or "just now" effectively mean the same thing. Now, not later.

Act 1, Scene 1:

The spring sun rises over a cloudless, temperate Johannesburg. The birds chirp in tall trees surrounding an average house on an average street in a quiet suburb. A South African father walks into his son's bedroom, noticing that the floor is barely visible and strewn with worn clothing.

FATHER

Good morning, my dear son! When are you going to pick up the clothes on the floor and put them in the wash?

SON

Good morning, Dad. Just now.

FATHER

OK, but before I get home this evening.

SON

OK, Dad!

The father leaves the room.

Fin

Please forgive the fabricated drama; it was for effect. The father understands that *anytime* between dawn and sunset, his son will clean up his room. This particular father-son relationship has shaped the father's perception of his son's response. Because "just now" is conditional upon the participants and the situation, the father understands that his son means "When I get to it." In order to deter another day of noncompliance, the father qualifies by requesting that the task conclude by day's end.

While researching this topic, the rabbit hole uncovered a concept called "linguistic relativity" or the "Sapir-Whorf hypothesis."[118] The concept relates to the structure of language and its effects on cognition—that is, the language you speak determines how you think. A society's language shapes its reality, and thus people who speak different languages have, via linguistic construction, different worldviews. There are words in some languages that have no direct literal counterpart in other languages, and since words confer meaning, descriptive words corresponding

117. This chapter's title is an homage to Raymond Carver, an American short-story writer and poet, and his collection titled *What We Talk about When We Talk about Love* (1981).

118. Sapir and Whorf never coauthored. Edward Sapir (anthropologist and linguist) first advanced the hypothesis in 1929, and Benjamin Whorf (his student) subsequently developed it.

to the untranslatable word will lack the linguistic nuance of the source language. Just ask any multilingual speaker.

The fascinating story of the Namibian Himba[119] tribe and their lack of a native word for the color blue elucidates this point well. Research shows that "blue" was one of the last color words to appear in many ancient languages, including Greek, Chinese, Japanese, and Hebrew. First, words for black and white (or dark and light) appeared in linguistics, then a word for red, then yellow and green, and finally blue. If a civilization or culture has no word for a color, in this case "blue," did members of the linguistically similar group even see (perceive) "blue"? The short answer is no, at least not as English speakers would perceive the color. The Himba borrowed the term "burou" for blue from their Herero neighbors, themselves borrowing it from the Afrikaans "blou." The research demonstrates that not having a name for something in a language, at the very least, stifles perception of the underlying concept.

The exercise had me thinking more deeply about temporal colloquialisms and expressions of clock time in different languages. I understand five languages but speak only one with (relative) mastery. Using this small subset as a foundation, I expanded my research to other languages and colloquial time-related[120] phrases in those languages.

Most modern languages incorporate similar rules relating to the expression of clock time. The cardinal hours on the clockface serve as primary reference points. Numerous languages reference half hours and quarter hours, relating both to the upcoming hour and the hour just passed. Most English speakers would use half past seven to mean 7:30, but in Dutch/Afrikaans[121] "half agt/acht"[122] means 7:30, even though literally translated it means "half eight," which is colloquially understood by Dutch/Afrikaans speakers as "half before eight."

In some languages, specific periods of a day have precise or semiprecise expressions, and the precision usually hinges on context. Noon, or more specifically meridiem (or meridian), means the literal midpoint of the sun's daily path but colloquially means 12 p.m. or "midday," which may extend from roughly an hour before noon to an hour (or more) after noon. Italians refer to noon as "mezzogiorno"—literally "middle of the day"—but it is also a reference to the geography of half of the country.[123] I do not know anyone who would refer to 10 a.m.[124] as midday, but 2 p.m.[125] could possibly satisfy the criterion. A similar pattern exists for midnight.

However, *afternoon*, sandwiched between morning and evening, stretches from 12:00:00 to about 18:00:00 (in military time, or 06:00:00 p.m. standard time), though there is no universal agreement on when afternoon ends and evening begins. Some speakers and locales would advocate afternoon as any time after noon but before sunset.

In India, Pakistan, Nepal, and Bangladesh, one refers to afternoon as "do-pahar." Literally meaning "second watch,"[126] it represents the second segment of the day. The twenty-four-hour day/night cycle is divided into four "watches"; hence midday coincides with the end of the second watch.

Biblical scholars and acolytes acknowledge time a little differently from standard definitions. Biblical evening begins at the sighting of three equivalently bright small stars similarly positioned in the evening sky. Discerning

119. Debi Roberson, Jules Davidoff, Ian R. L. Davies, and Laura R. Shapiro, "Color Categories: Evidence for the Cultural Relativity Hypothesis." *Cognitive Psychology* 50, no. 4 (June 2005): 378–411, https://doi.org/10.1016/j.cogpsych.2004.10.001.
There seems to be some controversy as to the exact tests administered to the Himba participants (and by whom). A BBC documentary based on the research appears to show blue/green color arrangements not specifically described in the aforementioned research paper. Nevertheless, in the paper the researchers themselves write: "There is a growing body of evidence that speakers of different languages, whose terminology or grammatical structure differ, encode, remember, and discriminate stimuli in different ways."

120. The word "time" in English is related to the Old English word "tima," from the Proto-Germanic tīmô—both related to the word for "tide," which means "a portion, extent, or space of time; an age, a season, a time, a while." (*Oxford English Dictionary*, 2nd ed., 1989).

121. Afrikaans is a Dutch-derived (Hollandic dialect) language spoken mostly in South Africa and Namibia and by some in Zambia, Swaziland, Botswana, Zimbabwe, and Malawi. It evolved from the original seventeenth-century Dutch brought to South Africa by early European settlers. More than 90 percent of Afrikaans originated from Dutch, with the remainder influenced by German and Khoisan.

122. The "g" or "ch" are guttural sounds, such as in the Scottish word "loch" (lake).

123. Southern Italy is known colloquially as Mezzogiorno in reference to the intensity and position of the sun at midday (as seen from the south). "Mezzogiorno," *Britannica*, https://www.britannica.com/place/Mezzogiorno.
Interesting fact: "time" in Italian is "tempo," a universally accepted term for the pace at which a passage of music is or should be played. In Italian, tempo may also mean "weather, period, day, stage, season, or tense."

124. a.m.: ante meridiem. Ante is the Latin prefix meaning "before."

125. p.m.: post meridiem. Post is the Latin prefix meaning "after."

126. Not to be confused with the time-telling device. Several ancient cultures divided the day or night (or both) into equal "watches" or periods.

the commencement of the biblical morning requires more effort. The Talmud[127] states that the span between dawn and sunrise is the amount of time it takes an average person to walk a distance of 4 "milin" (about 8,000 cubits or roughly 4 km[128]). According to most biblical decisors, an average person walks a distance of 1 mil in about eighteen minutes, and therefore seventy-two minutes is necessary to walk 4 milin. As a result, biblical dawn occurs seventy-two minutes before sunrise.[129]

IDIOMATIC TIME

Scores of idiomatic phrases exist in English and other languages referencing time and its elusiveness. Think of common phrases such as these:

- *It's a race against time*: meaning that one has little time to accomplish the task, and time may indeed run out before its completion;
- *All in good time*: meaning that one should be patient;
- *At all times, or at no time*: meaning either always or never;
- *From time to time*: meaning on occasion; and
- *It's about time*: expressing impatience at the event occurring only now.

INQUIRING ABOUT TIME

How do we inquire about the current time? In English, we would ask, "What time is it?" Examination of equivalent phrases in other languages produces interesting results. In Dutch/Afrikaans, the question is "Hoe laat is dit/het?" The phrase's literal interpretation is "How late is it?" German has the equivalent meaning for "Wie spät ist es?" Does this turn of phrase imply a philosophical approach to the lateness of the day, as if to indicate somehow that "earlier" is preferred? The Japanese ask "Ima, nan ji desu ka?" appending "right now" to the question, as in "What time is it right now?"

Some languages are tenseless in that they do not have a grammatical category for tense. They still talk about time but utilize descriptive words to establish temporal references, as is the case with Chinese. Some languages have two tenses[130] or three tenses—the traditional past, present, and future. The indigenous "Kalaw Lagaw Ya" language of Australia includes phrases for the remote past, the recent past, the today past, the present, the today / near future, and the remote future—six tenses in all.

NAMES OF THE DAYS OF THE WEEK

We use the standard seven-day week[131] in all forms of communication but spare little thought to consider its origin. If, like my children, you enjoy novels about ancient mythology,[132] then you will be familiar with the derivations. Most languages derive (if somewhat loosely) the names of the days of the week from Hellenistic astrology and its concomitant names for the classical "planets." Hellenistic astrology refers to the celestial spheres, ordered thusly: sun, moon, Mars, Mercury, Jupiter, Venus, and Saturn. A casual examination instantly explains where Sunday, Monday, and Saturday originate, but where do the names for Tuesday through Friday derive?

Tuesday comes from the Old English Tīwesdæg,[133] meaning "Tiw's day." Tiw, in Norse mythology, was a one-handed god. In Roman mythology, the mythological god of war, Mars, represented Tuesday. Wednesday hails from the Old English Wōdnesdæg, "the day of Woden" (in German mythology, or Odin in Norse mythology), itself derived from a Proto-Germanic word meaning "lord of frenzy." Its Latin counterpart is Mercury. Thursday is a dead ringer for Thor's day and in Afrikaans is called "Donderdag"—literally "Thunder day"—clearly referencing the mythical Norse god Thor. Similarly, Jupiter, the

127. Tractate (volume) Pesach/Passover, folio 94A.

128. The precise value of a cubit, and hence any measure dependent on it, is a subject of disagreement, since the cubit, by definition, is the length of arm from the elbow to the tip of the middle finger. Generally evaluated to be 18 inches, it could vary from 12 inches to as much as 24 inches, depending on the ancient cultural source.

129. Others support the view that it takes 22.5 minutes to walk a mil, and therefore ninety minutes to walk 4 milin. Accordingly, dawn occurs ninety minutes before sunrise. Regardless, biblical dawn refers to a specific point in the gradual transition from night to day. These transitions vary on the basis of latitude, altitude, and seasonality.

130. Curiously, English features only two tenses: past and present. Future tense isn't an "official" linguistic tense, because a statement such as "I will sleep" requires a modal (helping verb). Can, could, may, might, must, shall, should, will, and would are all modals.

131. According to ISO 8601 (the international standard dealing with dates and times), Monday is the first day of the week, followed by Tuesday, Wednesday, Thursday, Friday, Saturday, and Sunday. The biblical week traditionally starts with Sunday.

132. Especially of the Rick Riordan variety. In addition to other bestselling novels, Rick Riordan predominantly writes teenage-fiction book series that illuminate Greco-Roman, Egyptian, and Norse mythology, among other topics.

133. æ (uppercase Æ) is a character formed from the letters a and e. In modern Danish, Norwegian, and Icelandic (among others), it is an accepted letter of the alphabet.

mythical Roman god of thunder, represents Thursday. Last, Friday, from the Old English Frīgedæg, references the mythical Anglo-Saxon goddess Frige, corresponding to Venus or "Frigg's star."

Derivation for names of the months is no less intriguing. The Roman calendar derived names for the months from mythical Roman gods, festivals, leaders, and numbers.

January is the "month of Janus," the mythical Roman god of beginnings, transitions, and passages. Februa is the name of a Roman purification festival. March is for Mars, the mythical Roman god of war, since this month in the northern climes is seasonally propitious for starting wars. At one time, however, it was the first month in the Roman calendar. April derives from "Aprillis," which in turn connects to the Latin word "apero," meaning "second," when April was the second month. May stems from the name of the mythical Greek god Maia, nurturer of the earth—apropos for springtime. June is for Juno, mythical wife of Jupiter. July celebrates the birthday of Julius Caesar, following his assassination. August derives from Roman emperor Augustus. Its prior appellation was Sextilis when it slotted in as the sixth month in the Roman calendar. September, October, November, and December derive from the Latin cardinal numbers for seven (septem), eight (octō), nine (novem), and ten (decem). This of course was when December was the tenth month and March was the first, prior to the Julian calendar shifting the months around.

Temporal linguistics is a fascinating and exceptionally broad topic with many associated areas of study. I have only scratched the surface here.

TIME TRAVEL

"The other day, I happened to walk through it. I had a strange sensation. Not that time had passed, but that another me, a twin, was prowling around there—down to the smallest detail, and until the end of time—through what I had experienced over a very short period."

–Patrick Modiano, *L'Herbe des nuits*

This chapter is for my eight-year-old self, who may be reading it in an alternate timeline.

Would a book about time be complete without a section on time travel? Strictly speaking, we are all time travelers as we "move" through the ephemeral present.[134] However, at constant velocity and stable gravity, we are able to "move" into the future only one second at a time—that is, one second per second.

Einstein's well-tested theories of relativity provide for a limited type of time travel wherein relative acceleration/velocity or a strong gravity well provides these "time travelers" with a slower "objective" time experience and the ability to age more slowly compared to others not experiencing similar effects. Because the speed of light is invariant, time dilation (the difference in elapsed time measured by two reliable clocks) is real.[135]

Astronauts, pilots, airplane crew, and passengers are practical exemplars of those who are legitimate time travelers, albeit by picoseconds,[136] versus the general populace. Even at spacecraft velocities of 38,000 mph, time dilation effects are minuscule.[137]

With *exceptionally* vast resources, traveling into the more distant future is at least technically feasible.[138] The forces acting on the poor time traveler are another matter entirely, but assuming that the accelerative/gravitational effects of the exercise are survivable, one should emerge "in the future."

Scientists disagree on the possibility of practical backward time travel; for wont of a more precise definition, let us define it as "the ability to interact physically with a past event." Researchers continue to investigate the possibility of backward causality. Common sense would

134. See previous chapter on philosophy and science of time and (space).

135. Time dilation occurs in special relativity too, but with zero acceleration.

136. One trillionth of a second.

137. Voyager 1 (the fastest outward-bound spacecraft yet deployed) travels at about 38,027 mph or approximately 61,198 km/h. Light speed is often stated in meters per second or even in miles or kilometers per second. But such a comparison doesn't do justice to just how fast light speed is relative to even the swiftest of spaceships. Light speed is just over **670 million miles per hour** or **1.08 billion kilometers per hour**. Voyager 1 has covered about 1/418 of a light-year (about 14 billion miles as of late 2020) since launching more than forty years ago (September 5, 1977). It is currently traveling at roughly 1/18,000 the speed of light. At this rate, a journey to the nearest known star (other than the sun), Proxima Centauri, would take over 73,000 years. https://voyager.jpl.nasa.gov/mission/status/.

138. See previous discussion and illustration regarding the twin paradox in the previous chapter—"The Philosophy and Science of Time and (Space)."

normally stop us in our tracks. Einstein's theory of general relativity opens up the possibility for past time travel (perhaps via a rotating black hole), but then other physics models completely break down.

The famous "grandfather paradox" is often cited; a time traveler visits the past (assuming of course such a feat is possible) and murders his/her grandfather (rather gruesome, but it drives home the point—scientists can be quite morbid at times; consider also Schrödinger's alive/dead cat with its vial of poison). This parricide prevents the grandfather from fathering progeny, which would make the time traveler's existence impossible or at the very least nonintuitive.

Some physicists[139] believe we can avoid these temporal paradoxes through either the Novikov self-consistency principle[140] or a variation of the many-worlds interpretation.[141] In 1992, renowned theoretical physicist Stephen Hawking formulated the "chronology protection conjecture," suggesting that the fundamental laws of nature would rule out time travel.

Recent quantum-mechanical experiments have tried to test for self-consistency as it pertains to time, and the results are promising. Ostensibly, the probability of success in these quantum experiments is inversely proportional to the probability of photons interfering with themselves. In other words, the closer we get to a particle interacting with itself, the greater the chance the experiment will fail. In July 2020, scientists from Los Alamos National Laboratory demonstrated via simulation that the "butterfly effect"[142] does not exist for qubits (quantum bits) traveling from the past to the present in the quantum realm. On the basis of their results, time appears to be self-correcting.

Avoiding paradoxes is now also mathematically feasible. In September 2020, University of Queensland physicists provided sophisticated mathematical proofs confirming that free will remains intact and no paradox results when traveling from the present to the past. They proved logically that inconsistencies in what are known as closed time-like curves (where events from one point affect events prior and ahead) could be avoided. In short, time travel is mathematically possible, and you could *not* go back in time to kill your grandfather, because your very existence precludes it, and events will shape themselves around preventing the macabre outcome. Coresearcher Dr. Fabio Costa from the university's School of Mathematics and Physics provided an example: "In the Coronavirus patient zero example, you might try and stop patient zero from becoming infected, but in doing so you would catch the virus and become patient zero, or someone else would."

A few rays of hope exist for the optimistic would-be practical time traveler: wormholes and superluminal travel.

This hypothetical flat-earth conceptualization shows a wormhole connecting New York City to its antipode off the coast of Australia. Instead of traveling the roughly 20,020 km (approximately 12,440 miles) from New York City to the watery point in the middle of the Indian Ocean, one could use a theoretical wormhole to travel between the two points nearly instantaneously.

139. Allen Everett, "Time Travel Paradoxes, Path Integrals, and the Many Worlds Interpretation of Quantum Mechanics," *Physical Review D* 69, no. 12 (2004): 124023, https://doi.org/10.48550/arXiv.gr-qc/0410035.

140. The impossibility of creating time paradoxes within a closed time-like curve: any actions taken by a time traveler or by an object that travels back in time were part of history all along.

141. This would involve the time traveler traversing a different universe's history.

142. The term is closely associated with the work of Edward Lorenz, suggesting that a very small change in initial conditions may result in a significantly different outcome, such as the flapping of the wings of a butterfly influencing the formation of a tornado several weeks hence. The Los Alamos scientists understand time (at least in the quantum realm) to be self-correcting.

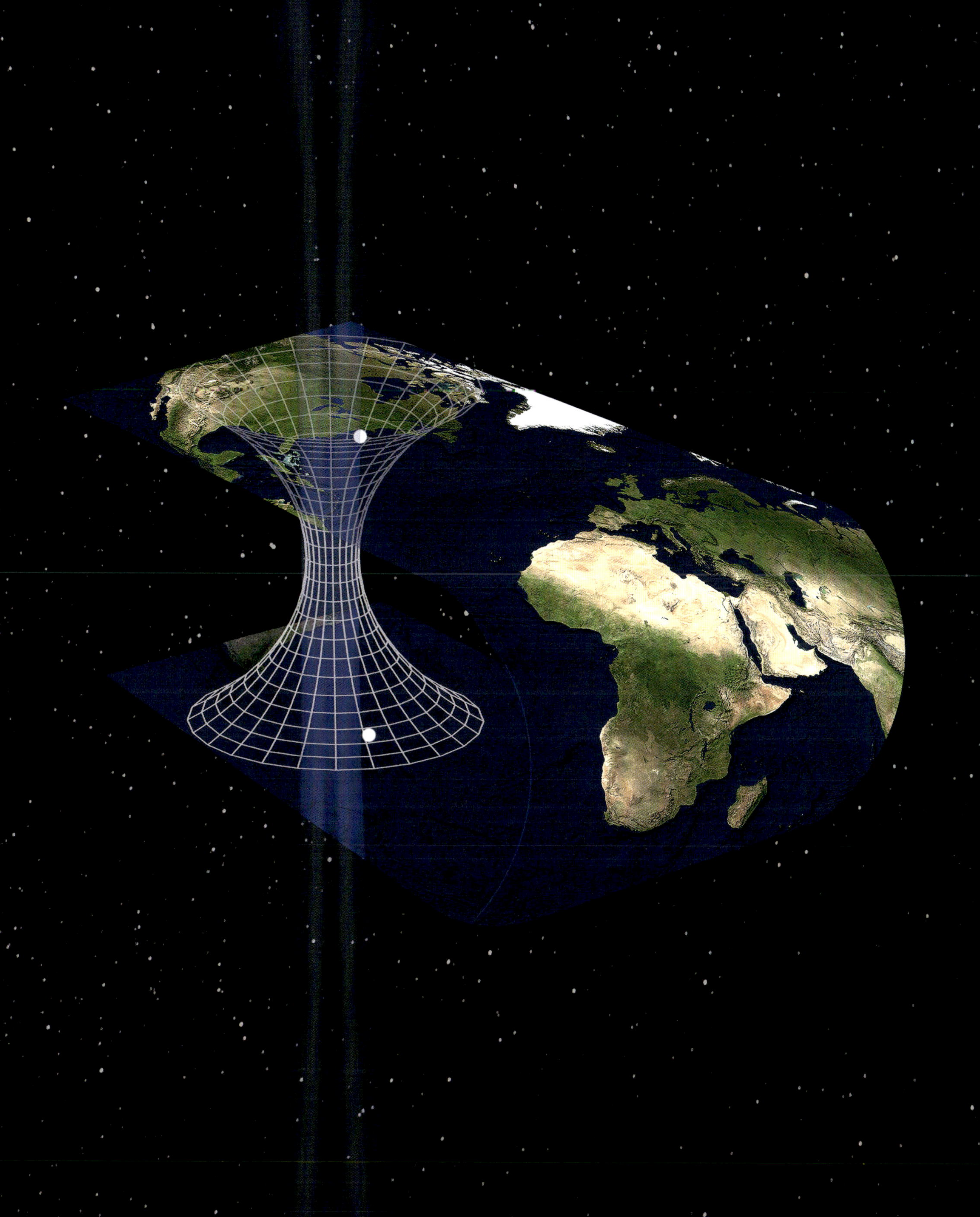

Wormholes are theoretical space-time tunnels that connect different regions of the universe—basically shortcuts, but on a cosmic scale—like folding a piece of paper on itself and punching a hole opposite the fold; previously distant areas of space (in this case the edges of the paper) are now adjacent. Mathematical proofs would suggest they do indeed exist. Unfortunately, we are yet to find any experimental evidence, and even if we do, we do not really know what would happen if we traveled through one. Furthermore, if we could construct a traversable wormhole, physicists believe that we would be able to go back in time only to the moment the "wormhole machine" initiated.

Faster-than-light (FTL or superluminal) travel is presumably off limits on the basis of our current understanding of causality.[143] We would also have to figure out how to acquire infinite mass to accelerate our shrinking non-zero-mass selves to light speed[144]—an impossibility on the basis of our current understanding of physics.[145]

Discussion of a variation of the Fermi paradox usually ends the fanciful dispute. Named for the Italian American physicist Enrico Fermi, the original paradox relates to the apparent contradiction between the lack of evidence for extraterrestrial civilizations and various high estimates for their probability. Simply stated, given the enormous number of galaxies, stars, and earthlike planets in the visible universe,[146] where is everybody else? So too with the absence of time travelers. If time travel exists at all within our universe's space-time continuum, then where are all the time travelers to validate it? Maybe time travel is possible, but time travelers either disguise their existence or use the technology too prudently for detection—no disappearing-in-a-flash DeLorean to give the game away.

Since H. G. Wells's 1895 story *The Time Machine*, the concept of mechanized time travel has fascinated the cultural zeitgeist. Historical records suggest that time travel stories are probably as old as humanity. As a species we are fascinated by time travel; one need look only to the vast number of science fiction books, TV shows, and movies with time travel as the central theme.

Time travel in science fiction falls into one of three categories: immutable timelines (what happened in the past will always be—the outcome is unchangeable), mutable timelines (a change to the past affects the future), and alternative histories or multiple timelines (historical changes manifest in new timelines with a different future). The previously discussed quantum-mechanical experimental results and mathematical proofs endorse the immutable timeline hypothesis, but study in this field is still in its infancy.

The aforementioned Hawking set up an experiment to entice future time travelers to visit. He published the invitations only after the event. No one showed up—at least not in this timeline. However, if you are reading this book in a future or alternative timeline where time travel is possible, the Hawking "reception for Time Travelers" was/is[147] held at "The University of Cambridge, Gonville & Caius College, Trinity Street, Cambridge. Location: 52°12′21″ N, 0°7′4.7″E, 12:00 UT 28 June 2009." If you find yourself in the neighborhood, do pop in; I am certain that late Professor Hawking would be delighted to meet you.

143. If light speed is the speed of causality, then exceeding this limit makes no sense.

144. A natural consequence of Einstein's famous $E=mc^2$ equation.

145. A non-zero-mass object accelerated to approach speeds close to the speed of light increases in mass, such that its inertia precludes it from further acceleration. Thus, an infinite amount of energy would be required for the particle to reach the speed of light, and the particle would have infinite mass.

146. On the basis of observations of relatively empty patches of sky (via the Hubble Space Telescope and others), as of 2020, scientists estimate that anywhere from a hundred billion to four trillion galaxies may exist in the observable universe alone. If each galaxy is assumed to be similar in size to our own Milky Way, with its hundred billion stars (at least—some estimates are much larger), the lower limit of stars in the observable universe is 100 billion × 100 billion = 1×10^{22}, or ten billion trillion stars. It stretches the limits of the imagination to think of how many planets and moons must exist if most stars have a solar system similar to our own.

147. Depending on your temporal perspective.

THE WRISTWATCH

"The chief beauty about time is that you cannot waste it in advance. The next year, the next day, the next hour are lying ready for you, as perfect, as unspoiled, as if you had never wasted or misapplied a single moment in all your life. You can turn over a new leaf every hour if you choose."

–Arnold Bennett

The concept of the wristwatch dates to the very earliest watches of the sixteenth century. In 1571, Elizabeth I of England reportedly received a wristwatch as a New Year's gift from Robert Dudley, the first Earl of Leicester. Described as an arm watch, its precise location on the arm has apparently been lost to history.[148]

Breguet's meticulously preserved archives include the "register of commissions." Page 29 of the register shows that on June 8, 1810, the Queen of Naples (Maria Carolina, wife of King Ferdinand IV) commissioned two timepieces. One of these timepieces, watch number 2639, would become the world's first known extant example of a wristwatch.[149]

Until the early twentieth century, wristwatches were almost exclusively a woman's accessory. Men initially found these timepiece oddities to be effeminate, but the necessity for easy-access timekeeping on the nineteenth-century battlefield quickly changed this perception.

WHY OWN A WATCH?

The reasons for owning a watch are manifold. Survey a dozen people and you may get a dozen different answers. Nevertheless, the majority of watch owners' reasons for ownership fall into a few well-defined categories.

(1) MEASURING TIME

Simply knowing the time is highly useful in the conduct of one's everyday affairs, and the act of telling time is one of man's most ancient activities. Lunar calendars were among the first types of time measurement to appear and are still extant in many parts of the world. Early-twenty-first-century humans are no less fascinated with time measurement; the surfeit of complicated mechanical timepieces bears this out. Take for instance the Vacheron Constantin Reference 57260, the most complicated (and perhaps the most expensive)

148. *Journal of the British Archaeological Association* 20 (1864).
149. Montres Breguet SA.

watch ever produced.[150] Three distinguished watchmakers created its fifty-seven complications over a period of eight years.[151]

Released in 2015, it commemorated Vacheron Constantin's 260th anniversary. Among its features are six time measurement functions, seven perpetual-calendar functions, eight Hebrew calendar functions, and nine astronomical-calendar functions.

(2) HANDS-FREE TIME TELLING AND ALWAYS ON

In today's hurried world, always knowing the time is for most a necessity. While smartphones also reliably indicate time, these devices require retrieval from one's pocket or bag. A wristwatch, on the other hand (yes, pun intended), is designed to be accessed at a moment's notice. In fact, the popularization of the wristwatch occurred because World War I soldiers needed to instantly access the time when implementing coordinated battle tactics. Searching for one's pocket watch in the heat of battle required too many steps and distracted the soldier from the immediate events on the battlefield. Moreover, the instantaneous time-telling abilities of the wristwatch require no powering-on of a device and no constant recharging, with the obvious exception of the smartwatch.

Urban Jürgensen (Ref. 1142PT, MSRP: $46,600) with 42 mm platinum case and silver dial. The company, originally founded in 1773, still maintains the extensive degree of artisanship of its founder.

Urban Jürgensen, son of Jürgen Jürgensen, followed his father's profession and apprenticed under watchmaking luminaries Abraham-Louis Breguet and John Arnold. The thermally blued steel hands are handcrafted. They adorn the basketweave-patterned guilloche dial. The hour hand circle is fashioned from gold, utilizing diamond polishing. The dial is a single piece of silver.

The hand-wound Urban Jürgensen, Calibre P4, powers this unassuming watch. The movement, as meticulously finished as the dial, features hacking seconds and a sixty-hour power reserve.

150. As of this writing.

151. In addition to the standard hours, minutes, and seconds, the fortunate owner of the 57260 is also able to observe a twelve-hour time zone; a second hours-and-minutes time zone; a twenty-four-city display; a day/night indication; a Gregorian perpetual calendar with day, month, and retrograde date; a leap year and four-year-cycle display; the number of the day of the week; a week to view; a Hebrew perpetual calendar and nineteen-year cycle; the Hebrew day, month, and date; the Hebrew secular calendar; the Hebrew century, decade, and year; the age of the Hebrew year (twelve or thirteen months); the golden number (nineteen-year cycle); the seasons, equinoxes, solstices, and signs of the zodiac indicated by the hand on the sun; a star chart representing the sky above the owner's city; the sidereal-time hours and minutes; the equation of time; the sunrise, sunset, length-of-day, and length-of-night times for the owner's city; the moon phases and age (accurate to one day in 1,027 years); the date of Yom Kippur; and a chronograph and split-seconds chronograph with hours and minutes counter.

URBAN JÜRGENSEN
SWISS

(3) HAVING A MORE INTIMATE RELATIONSHIP WITH TIME

Time has long been a major subject of study in religion, philosophy, and science. Two contrasting viewpoints on time divide many prominent philosophers.

The Newtonian view is that time is part of the fundamental structure of the universe—a dimension independent of events, in which events occur in sequence. The opposing Leibniz-Kant view is that time does not refer to any kind of container that events and objects move through, nor to any entity that flows, but that it is instead part of a fundamental intellectual structure (together with space and number), within which humans sequence and compare events. Refer to the chapter on the "Philosophy of Time (and Space)" for additional thoughts.

We all have a unique relationship with time, and it interweaves intimately with our everyday existence. Celebrations of events such as birthdays and anniversaries are a testament to our fascination with the need to celebrate time's passage.

Heuer Carrera 1158 CHN (ca. 1972). The tonneau shape (cushion/oval hybrid) is an exceptionally popular design among vintage Heuer enthusiasts, and only a few examples (in gold) come up for auction each year.

Jack Heuer (the great-grandson of the eponymous founder) has spoken publicly about this being his favorite watch. This timepiece features Heuer's first automatic in-house chronograph (introduced in 1969). Reference 1158 CHN watches were famously gifted by Heuer to the Formula 1 Ferrari team drivers from 1971 to 1979—and they didn't end up in a drawer either. Numerous photographs from the time show the drivers wearing these durable timepieces on track.

TACHY
CARRERA
HEUER
AUTOMATIC
CHRONOGRAPH
SWISS

(4) AS A CONVERSATION PIECE

What is that you are wearing? This question is a great conversation starter, and one the horophile[152] is eager to address.

There is plenty to talk about too; from multicomplications, with their exquisite intermeshing of gears, pinions, and cams, to unique materials, brand heritage, or fashion-forward style. For many, being noticed with one's timepiece is a badge of honor. For others, discretion is the order of the day, and only the true aficionado will ever notice their timepiece.

152. A person obsessed with timekeeping devices and watchmaking.

"Excuse me! What is that timepiece that you are wearing?" The Romain Jerome Steampunk Titanic Auto 100th Anniversary (ref. RJ.T.AU.SP.002.02, MSRP: $16,950) launched in 2012 on the one hundredth anniversary of the sinking of the *Titanic*. It is 50 mm wide with signature rusted bezels (containing actual melted and reprocessed metal from the doomed ocean liner).

(5) AS FUNCTIONAL ARTISTIC SCULPTURE

Artistic works have existed for almost as long as humankind has: from early prehistoric art to contemporary design. One early sense of the definition of art relates closely to the older Latin meaning, which roughly translates to skill or craft.[153]

Certainly, at the uppermost levels of watchmaking, this artisanal craft knows no equal. The wristwatch allows its wearer to behold its artistic form—be it the specific shape of the watch; or its texture, color, or style; or perhaps the interaction of the harmony of the individual components to the whole. As a three-dimensional form, the wristwatch doubles as wearable sculpture.

Founded in 2002 by Morten Linde and Jorn Werdelin, Linde Werdelin is a watch and instrument company. The Spido and Oktopus families make up their collections with the chronograph Spido watches designated SpidoSpeeds and the nonchrono versions dubbed SpidoLites.

Shown here is the highly sculptural skeletonized case of the SpidoSpeed in rose gold (Ref. SPS.G.RG.A, MSRP: $38,500). Movement is the automatic LW 03 (Concepto Calibre 2251). The detailed case skeletonization results in a timepiece weighing a "scant" 126 grams, despite the 44 mm diameter case consisting almost entirely of rather dense gold.

153. Associated with words such as artisan and derived English words such as artifact.

LINDE WERDELIN

De Bethune's signature DB28 titanium wristwatch featuring a **spherical moon phase** (Ref. DB28TI, ca. 2011, last known MSRP: 89,400 CHF).

De Bethune's patented "floating lugs" provide ultimate wrist comfort for this architectural timepiece. The mirror-polished (platinum and blued steel) spherical moon phase is another technical marvel of this rare and beautiful watch. The watch houses Calibre DB 2115 (28,800 vph, 144-hour power reserve, thirty-six jewels, 293 parts). Diameter is 42.6 mm; thickness 9.30 mm.

This model won GPHG's 2011 much-coveted "l'Aiguille d'Or" award (for the best watch of the year), which is akin to being named the year's best picture at the Oscars.

Mechanical sculpture. KIVA Halo Chronograph in brushed and high-polish grade 5 titanium (Ref: 272, MSRP: $7,650, ca. 2010).

The multipart CNC-manufactured case features two concave scalloped shock-absorbing chronograph-pusher housings that flank the crown on the right case band, as well as a concave scalloped shock-absorption device (mirroring the chronograph housing curvature) that flanks the left case band. In addition to these appurtenances' deliberate sculptural qualities and unique polished titanium sheen, they are highly functional and designed to protect the watch and movement from vibrations and unintended knocks.

The crown features two polished sipes and a microknurled finish, providing pleasant tactility and effortless control. Fluted-gadroon chronograph pushers assist start-stop/reset functions and protect against unintended finger slippage.[154] Magnified date at three o'clock provides easy date read-off for the visually impaired.

154. I personally designed the KIVA Chronograph, so in terms of creative intent, I can be relatively certain of the designer's objectives.

KIVA
ALL TITANIUM
SAPPHIRE CRYSTAL
Tachymeter
MON 24
AUTOMATIC
300M/1000FT
SWISS MADE

(6) STYLE OR FASHION, SELF-EXPRESSION, OR TO ACCESSORIZE AN OUTFIT

Personal expression takes many forms, including the way in which one dresses. A watch is usually a very personal choice. It may reflect one's desire to mimic the pursuits of the men and women displayed in the many and varied advertising campaigns of the watch brands. Perhaps the wearer associates with the aviator, the diver, or the explorer. He or she may admire a celebrity endorsing a particular brand. Some may be enamored with the fashionable or avant-garde design of the timepiece. Those who are fortunate to own more than one timepiece may accessorize their particular outfit with the most appropriate timepiece from their collection.

The classic styling of Patek Philippe's 1463 (produced from 1940 to 1965), shown here in yellow gold with applied arrow markers.

The 1463's claim to fame is that it was Patek's first "waterproof" chronograph. Previous chronographs were fitted with snap-backs—this one features a screw-back. Most examples in the estimated 750-piece run left the factory in yellow gold.

The 1463 embodies numerous vintage Patek Philippe styling cues and is a superb example of mid-twentieth-century Patek. As of the end of 2020, excellent examples were fetching around $350,000—significantly undervalued (in relation to its better-performing peer group) in my opinion for a water-resistant chronograph "first."

PATEK PHILIPPE
GENÈVE

(7) FUNCTIONAL BACKUP

Early-twentieth-century military personnel, aviators, submariners, and explorers did not have the luxury of twenty-first-century technology to provide reliable and precise time measurement. For these people, a rugged and functional mechanical timepiece was a necessity, be it for strategic or tactical coordination, navigational computation, dive-time measurement, or environmental resiliency. Modern dive computers, GPS devices, altimeters, and other equipment assist today's adventurers in their pursuits, but the specialized wristwatch accompanies many of these individuals in their activities and provides a reliable backup, should twenty-first-century technology fail them in their time of need.

Mühle-Glashütte Sea-Timer BlackMotion Ref. M1-41-83-NB in stainless steel with ultra-hard black titanium carbide coating. With 300-meter water resistance. MSRP: $2,650.

NAUTISCHE INSTRUMENTE
MÜHLE
GLASHÜTTE/SA.
Sea-Timer
BlackMotion
30 atm
MADE IN
GERMANY
14
10
20
30
40
50

(8) STATUS OR "YOU CAN'T DRIVE YOUR CAR INTO THE BOARDROOM"

You cannot drive your car into the boardroom! Your supercar or luxury sedan may broadcast your status to the world while driving, but in the boardroom, it is hidden from public view. Many individuals wear a suit or other stylish attire, but exuding one's status (if that is one's desire) in the workplace is not so simple. Enter the luxury timepiece. For decades, many up-and-comers on Wall Street were (and many still are) infatuated with owning a Rolex. Owning a luxury timepiece such as a Rolex or Patek Philippe signaled to the world that you had made it; you were part of an elite club. Scores of brands exist in the world of Haute Horlogerie, each one confirming on its owner a pedigree of perceived status and taste.

An otherwise unremarkable sports watch with a hidden tourbillon. To the uninitiated, this timepiece may easily fly under the radar, but to the horological cognoscenti, this ultrarare $185,000 limited-edition Laurent Ferrier Tourbillon Grand Sport released in 2019—one of only twelve—is the ultimate expression of boardroom bragging rights.

On the underside is an unobstructed view of Laurent Ferrier's calibre LF 619.01 with double spiral tourbillon.

LAURENT FERRIE
GENEVE
TOURBILLON GRAND SPORT
SWISS

(9) TIME MANAGEMENT

Recording or at least being cognizant of the time spent on specific activities increases one's effectiveness, efficiency, and productivity. At the core of an effective time-management system is methodical time tracking, and the wristwatch can be indispensable in a variety of measurement tasks. Because of its portability, the wristwatch is always with its wearer whether he or she is walking, running, driving, sitting, lounging, climbing, diving, flying, or involved in any other of one's myriad daily activities.

Casio G-Shock Full Metal 5000 (Ref. GMW-B5000D-1, ca. 2018, MSRP: $550) in all-metal construction was inspired by Casio's first G-Shock model from 1983. Its designer, Kikuo Ibe, wanted to construct an unbreakable watch based on the triple-10 philosophy: water resistant to 10 bar (100 m), a minimum 10 years of battery life, and, most importantly, the ability to survive a 10-meter fall.

The watch has since sold more than a hundred million units. At least two hundred new variations are available for purchase. G-Shock is an abbreviation of Gravitational Shock. Given its incredible resistance to damage, it is no wonder that the G-Shock continues to be popular with all manner of emergency service and military personnel.

PROTECTION
CASIO
TOUGH SOLAR
SHOCK RESIST
◀ SPLIT·RESET
LIGHT ▶
PS
WE
10.23
P
3:05 44
ADJUST
MODE
SET [–]
SET [+]
WATER RESIST
20BAR
RECEIVING ▶
START·STOP ▶
Bluetooth
MULTI BAND 6
G-SHOCK

(10) NOSTALGIA

Happy personal associations with the past drive us to seek out reminders of these pleasant events. Perhaps you recall the well-worn Omega Speedmaster that your father wore when you were a child. Perhaps you are just enamored with watches produced during a simpler time. The watch industry is certainly happy to oblige with modern reeditions of distinctive timepieces from the 1920s through the 1970s. New editions evocative of the 1980s are just starting to make it to the marketplace. If the modern-day equivalent is not your speed, there is a hugely successful and rapidly growing vintage watch market to satisfy your stylistic memories of the past.

Breguet Type XX Transatlantique, Ref. 3827PT/32/9W6 (ca. 1998) in platinum, with salmon dial and exhibition case back. This is the third-generation Type 20, now with the Roman moniker XX (instead of 20, to prevent other brands from using the same designation and facilitating Breguet's ability to trademark the collection).

Breguet has been involved in aviation for more than a century. First introduced in 1994, the Type XX features a 39 mm case diameter, Calibre 582/1, with twenty-five jewels and a flyback chronograph.

The French air force commissioned 2,000 Type 20 Generation 1 models in 1954. They sold concurrently with Breguet's similar-looking civilian models. The illustrated piece is limited to just ninety examples. Unlike other Type 20s, with printed numerals, the limited-edition platinum version comes with hand-applied raised Arabic markers.

Breguet
SWISS MADE

(11) MARKING AN OCCASION

The purchase or gifting of a timepiece to mark the occasion of a promotion, retirement, birth, engagement, marriage, anniversary, graduation, contract signing, or other important or significant event is a well-established tradition. Many of these keepsakes carry engravings with statements of recognition, affection, or accomplishment. Because a fine timepiece is likely to outlast its owner, it is often bestowed to one's beneficiaries.

One of the most famous wristwatch bequests is the 1968 Rolex Daytona sold by Phillips auction house in October 2017. Including buyer's premium, the record-setting sale of $17,752,500 arrived after twelve minutes of intense bidding. It smashed the previous record-setting amount of $11 million set by a Patek Philippe 1518 from November 2016, which was also sold by Phillips (in association with Bacs & Russo). Engraved on the case back of this highly sought-after Rolex timepiece are the words "Drive Carefully Me." The inscribed sentiment was from actor Joanne Woodward to her racing-driver husband and fellow actor, Paul Newman.

The view from the underside of the Vacheron Constantin 87172/000G-9301 (MSRP: $24,700) in 18K white gold, engraved by its benefactor: "To JT, Love LM."

GENEVE SWISS
22K
To JT, Love LM

(12) AS A MAN'S ONLY PIECE OF JEWELRY

Women are privileged to be able to wear all manner of jewelry, from engagement rings to nose rings, belly rings to toe rings, and chokers to ankle bracelets. Many men feel uncomfortable sporting any kind of jewelry at all, beyond a simple wedding ring. For them, a watch serves a dual function—timekeeper and jewelry.

The A. Lange & Söhne Saxonia Automatic in white gold Ref. 380.026. Discreet 38.5 mm case diameter; downturned lugs with alligator leather strap. Seventy-two-hour power reserve. An unobtrusive choice that is always in style. MSRP: $27,000 (ca. 2012).

A. LANGE & SÖHNE
GLASHÜTTE I/SA
MADE IN GERMANY

(13) INVESTMENT

Most watches are not particularly good investment assets. Like most forms of art, the vicissitudes of a fickle marketplace and the supply-and-demand equation determine the price of watches. Ironically, watches that typically proved unpopular at launch, thus engendering exceedingly small production runs, end up decades later commanding huge premiums arising from their rarity.

Several brands drive huge premiums in the vintage auction environment, most notably Patek Philippe and Rolex with their significant and loyal following of connoisseurs and collectors.

Recent auctions of complicated Patek Philippe timepieces have commanded million-dollar or even multimillion-dollar prices for watches that would have sold for $10,000 or less just several decades prior.

Patek Philippe 2499 in pink (rose) gold (ca. 1955–1966). The rarest of birds and a historically resilient investment—fewer than ten examples exist in this precious metal. Horophiles simply call it the "twenty-four ninety-nine." **It is possibly the single most collectible wristwatch in the world.** Only 349 examples left the factory over its thirty-five-year run (across all four versions). Version 2, shown here, is the scarcest and most sought after, with its round chronograph pushers and tachymetric scale.

Very few examples of the 2499 ever make it to auction. A platinum 2499 owned by renowned musician Eric Clapton sold for $3.65 million in 2012. Christie's sold a magnificent example of the pink gold reference for CHF 1,985,000 in 2013. In June 2020, Jean-Claude Biver's personal 2499 (in yellow gold) sold for $2.74 million. If you are fortunate enough to find one in pink gold available today, expect to pay upward of $4 million (and now probably upward of $6 million, following the "Nevadian sale").

BASE 1 MILE
SUN
JUL
PATEK PHILIPPE
GENÈVE

THE WATCH AND ITS MOVEMENT

> "For the Present is the point at which time touches eternity."
>
> – C. S. Lewis, *The Screwtape Letters*

If the watchcase is the body of the timepiece and the dial is its face, then the movement would undoubtedly be its organs.

Like the human with his or her regularly beating heart, the movement contains a precisely engineered balance spring that gently twists back and forth many thousands of times an hour. In fact, over a year, most modern mechanical wristwatches beat approximately two hundred million times. If you service your watch routinely (every five years is the recommended interval), your watch will have experienced *one billion beats*. Remarkable!

THE MAINSPRING

The timepiece requires power to operate. In a mechanical wristwatch, the mainspring provides this power; a tightly coiled flat steel spring (commonly spring steel) acquires its energy from the winding of the crown or the rotation of an attached rotor (found in self-winding watches).

Shown at about five times actual size

Just as the spring in a windup toy car winds when the car is pulled backward along the surface, so too does the crown (or rotor) wind the mainspring (housed within the barrel drum and secured by a barrel cover on its underside).

The ultrathin 0.05 mm to 0.2 mm thick mainspring is 0.6 to 3.5 mm tall and about 200 to 300 mm long (about a foot) when uncoiled. The mainspring would rapidly uncoil its power potential were it not securely attached to the going train (see below). The relatively enormous torque provided by the mainspring gradually releases via the meted-out oscillations of the balance wheel found at the opposite end (to the mainspring barrel) of a series of wheels and pinions—known as the going train, wheel train, or gear train. The weighted balance wheel receives an impulse from the pallet lever, actuated in turn by the escape wheel.

Thus, the mainspring provides power through a series of large and small wheels until the power transfers to the wheel called the escape wheel. This power releases in precise units, regulated by the oscillations of the balance wheel.

In the case of a quartz timepiece (see below), a battery or some other source of stored power would generally replace the mainspring.

THE BALANCE

The balance wheel oscillates back and forth, which is analogous to the swing of a pendulum. It returns to center when the spiral or helical torsion balance spring (to which it is connected) twists back. The balance spring is essential to the workings of the balance wheel. Together they form a harmonic oscillator (the balance) that oscillates at a specific period or beat, ensuring the extreme accuracy of the timekeeper.

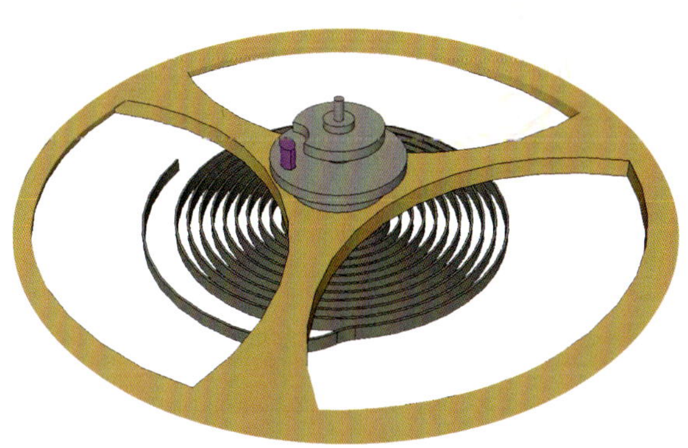

When first introduced to pocket watches by Robert Hooke or Christiaan Huygens (or both) ca. 1660, the balance spring instantly transformed these devices into useful timekeeping instruments.

Temperature change is one of the balance spring's greatest antagonists. Modern materials alloys science has innovated alloys to improve thermal expansion coefficients; these alloys morphed from Invar to Elinvar and finally to Nivarox in 1933.

Nivarox is still in use today and dominates as the leading choice for the majority of balance springs. The company Nivarox-FAR is now part of the Swatch group, and, after the 1970s quartz crisis, the only manufacturer of the "Nicht Variable Oxydfest" (**NiVarOx**) balance spring—an alloy consisting of iron, nickel, cobalt, chromium, titanium, and beryllium. In recent decades, other watch groups and manufacturers have sought substitutes to the "monopolistic" Nivarox supply.

Several brands, including Bovet, H. Moser & Cie, Vaucher, A. Lange & Söhne, and Jaeger-LeCoultre, have all developed alternatives. In the first few years of the 2000s, Patek Philippe, Rolex, and Breguet collaborated with the Swiss Center for Electronics and Microtechnology (CSEM) to develop a silicon hairspring that was light and amagnetic and demonstrated precise forms, enhanced isochronism, and superior thermal compensation. Moreover, silicon does not require lubrication and is etched (using DRIE: deep reactive ion etching) with extreme precision. Notably, Ulysse Nardin also developed a silicon hairspring and was actually first to unveil it to the public in 2001, encased in the "Freak" timepiece. Numerous brands now utilize silicon hairsprings.[155]

The use of lubricant-free silicon for the balance spring has allowed manufacturers to extend their warranties and concomitant service intervals for their timepieces. Silicon hairsprings (another name for balance springs) have their detractors. They are brittle, difficult to repair (except by specialists), and, for some, soulless. Watches from the nineteenth century are still repairable today with much the same techniques employed one hundred years ago. Will silicon have the same serviceability one hundred years hence?

In a quartz timepiece, the quartz crystal (see below) replaces the balance.

THE ESCAPEMENT

Together, the escape wheel and pallet fork (or lever) make up the escapement.

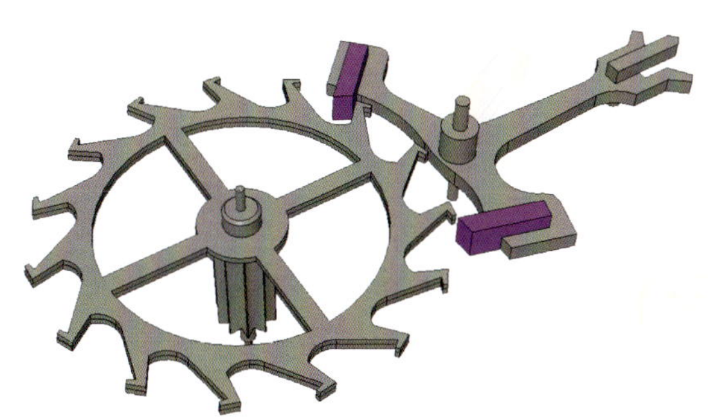

When knocked by the impulse pin on the balance wheel, the trident-shaped pallet fork moves side to side, swinging its other end (the pallets) back and forth. The pallet fork and pallets combine to form the "lever," shaped

155. The development of novel hairsprings continues. In early 2019, TAG Heuer introduced a carbon composite balance spring as a further alternative to Nivarox.

like the letter T. The pallets are two rectangular jewels located at either end of the crossbar of the T. These jewels[156] lock and unlock the escape wheel.

One oscillation of the balance—that is, one back and forth movement—will shift the escape wheel rotationally by one tooth. In a 21,600 vph[157] movement, the escape wheel usually contains fifteen teeth and moves one full rotation every five seconds.[158] The frequency of the movement largely determines the escape wheel tooth count.

THE GOING TRAIN

The mainspring lies at one "end" of the movement, and the balance at the other end. Between them is a system of gears called the going train.

Watchmakers refer to the large gears as wheels and the smaller gears as pinions. The shaft connecting to a wheel is called the arbor. The movement plates keep all the wheels in place. In order to ensure that the power supplied by the mainspring lasts the entire day (or even longer), a series of gears scales up the speed of rotation of the mainspring (much like a bicycle's or car's gears). The ratio of gears (based on the number of teeth of collocated wheels) properly divides the time into measured units of seconds, minutes, and hours, represented by the concomitant display hands connected to the relevant arbors. The going-train wheels experience the most wear and tear in the movement, since they bear the constant torque of the mainspring.

The mainspring barrel (described above) attaches to (or incorporates) the first or great wheel. The pinion of the great wheel turns the center wheel (once per hour), and its arbor drives the cannon pinion, carrying the minute hand. The hour wheel, connected to the cannon pinion, rotates once every twelve rotations of the cannon pinion. The third wheel drives the pinion of the fourth

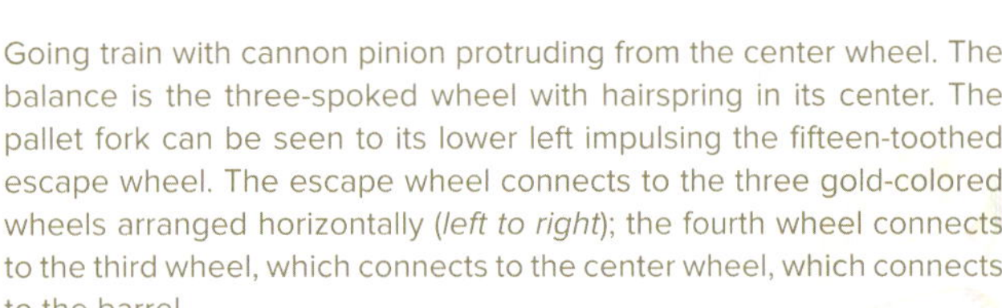

Going train with cannon pinion protruding from the center wheel. The balance is the three-spoked wheel with hairspring in its center. The pallet fork can be seen to its lower left impulsing the fifteen-toothed escape wheel. The escape wheel connects to the three gold-colored wheels arranged horizontally (*left to right*); the fourth wheel connects to the third wheel, which connects to the center wheel, which connects to the barrel.

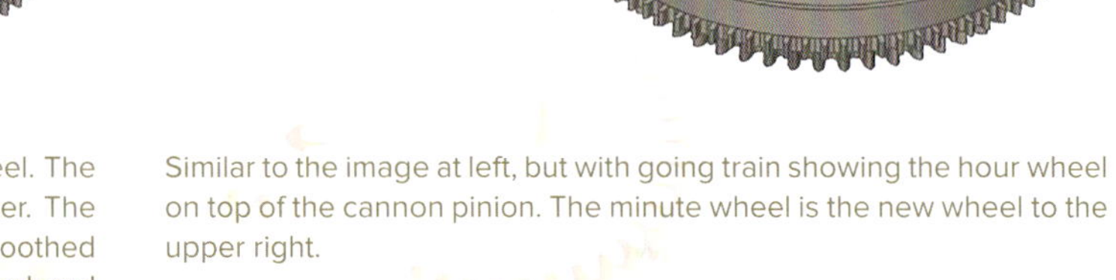

Similar to the image at left, but with going train showing the hour wheel on top of the cannon pinion. The minute wheel is the new wheel to the upper right.

156. Ruby, usually synthetic corundum.

157. Vibrations per hour or when displayed as bph, beats per hour.

158. 21,600 (divided by) 60 minutes (divided by) 60 seconds (divided by) 2 oscillations per "tick" = 3—which divides into 15 five times; for a 28,800 vph movement, the escape wheel usually features twenty teeth but vibrates with eight oscillations—four back-and-forth movements per second—and it also moves the escape wheel once every five seconds.

wheel, which contains the arbor of the second hand. The fourth wheel also turns the escape wheel pinion, which drives the escape wheel—releasing one tooth at a time per oscillation of the balance.

Fully wound, the mainspring—or mainsprings (in watches with longer power reserves)—produce many hours of energy to the movement—from as few as thirty-six hours to as many as fifty days—and in the case of the automatic or self-winding watch, normal human movement is sufficient to keep it wound almost indefinitely.

Watch movements may contain hundreds or even thousands of almost microscopic parts, each painstakingly designed to fit and collaborate with each other. Well maintained, a mechanical timepiece will provide a lifetime (or possibly lifetimes) of reliable service.

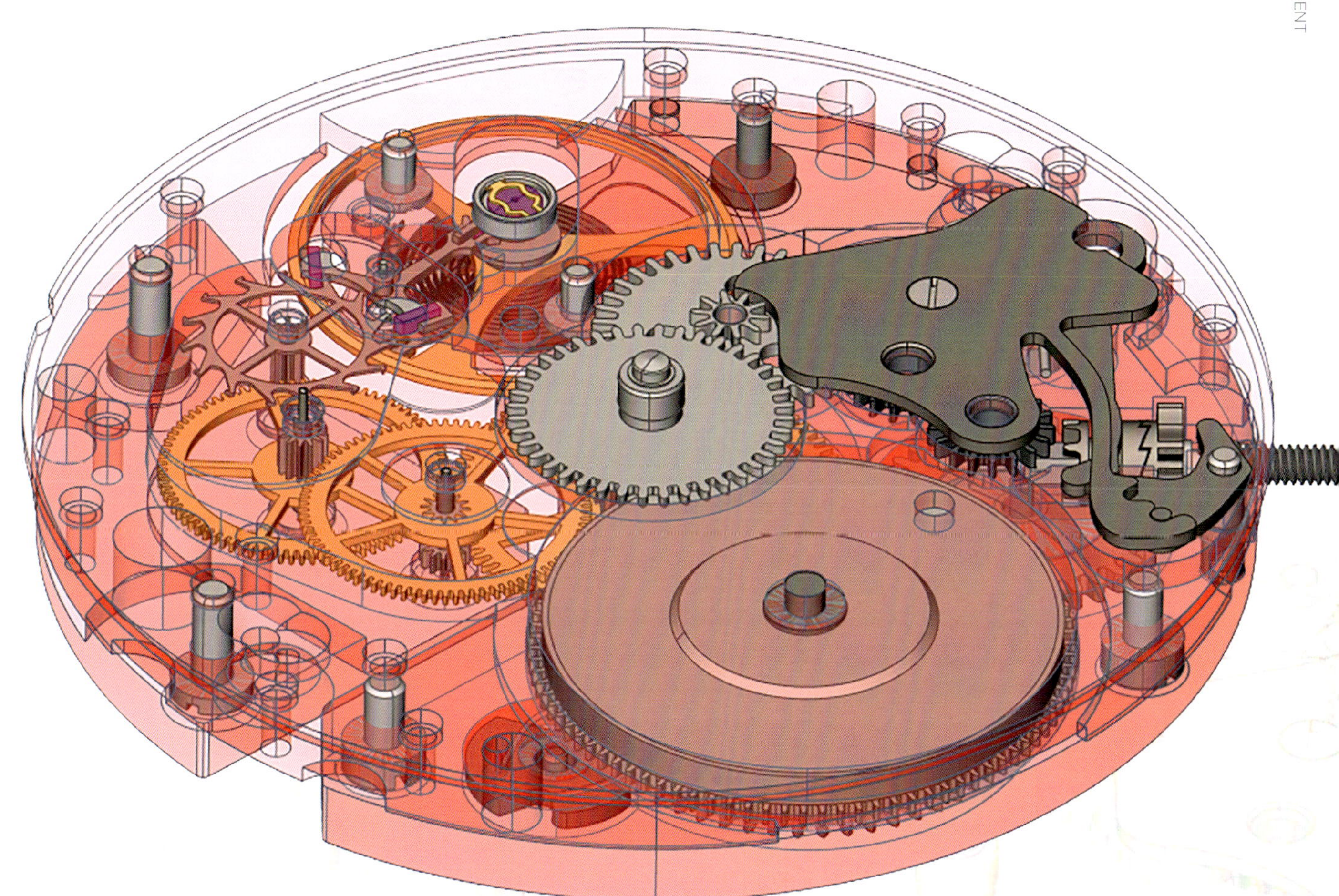

A fully assembled transparency-layered CAD drawing of the Unitas 6497-1 movement—a relatively simple yet reliable workhorse movement. The movement is 16½‴,[159] fully winds to offer a forty-six-hour power reserve, contains seventeen jewels, and beats at 18,000 alternations (a complete back-and-forth oscillation) per hour. The hour wheel atop the center wheel and cannon pinion is in the center of the image. The large mainspring barrel is located toward the lower right. The winding stem connects at the far right.

159. Before the metric system, the "ligne" or Paris line defined a unit of length. A meter is 443.296 French lines. Thus, 1 ligne is about 2.2558 mm. Lignes are designated with the triple prime (‴), and 12 lignes make up 1 French inch or pouce. The ligne is still in use today by European watchmakers. For instance, 16½ lignes or 16½‴ is called size 9 and is 37.221 mm in diameter.

PARTS OF A MOVEMENT

A: Balance
B: Incabloc[160]
C: Hairspring
D: Pallet fork
E: Screw for train bridge
F: Pallet jewel
G: Escape wheel
H: Subsidiary second hand (above dial)
I: Fourth wheel (alt. "second" wheel)
J: Third wheel
K: Center wheel
L: Dial numeral (on dial)
M: Minute track (on dial)
N: Yoke
O: Minute wheel
P: Hour hand (above dial)
Q: Intermediate setting wheel
R: Minute hand (above dial)
S: Crown wheel
T: Sliding pinion
U: Winding stem
V: Setting wheel
W: Driver cannon pinion
X: Hour wheel
Y: Setting lever
Z: Ratchet wheel
AA: Mainspring barrel (first wheel)
AB: Mainspring
AC: Click

160. Incabloc is a shock protection system. It is the trade name for a spring-mounted, jewel-bearing system supporting the delicate balance wheel (prone to shock and damage, should the watch suffer a knock or drop). Similar trademarked systems exist with the same overall purpose, such as Paraflex, Etachoc, Kif, Diashock, and Parashock.

PIECING IT ALL TOGETHER

If the aforementioned description has you thoroughly (or perhaps only moderately) confused, I have illustrated additional technical drawings to assist understanding regarding part names and their numerous interactions.

First up is an image of "the Escapement." The impulse pin (partially obscured) is the corundum (ruby) jewel embedded in the safety roller that slides with the motion of the roller (itself propelled by the motion of the oscillating balance wheel—not shown). The white arrow indicates motion of the roller (although it does go in the opposite direction, during the reverse swing of the balance). The guard pin (the elongated middle prong of the "trident"-shaped fork) prevents the fork from inadvertently "jumping," preventing the impulse pin from entering the fork slot (the gap between the horns of the fork). Simply put, it keeps the fork in its proper place. Below the main image of the escapement is an inset (orange) circle that displays the lever jewel planes. These designations will be useful in the "tick tick tick" image explicating the nature of the ticking sounds.

The double-page image of the balance and shock protection assembly illustrates how the shock protection system works. The cap jewel and balance staff jewel nestle on top and within a mobile bushing inside the endpiece, below a specially shaped shock protection spring. Should the balance undergo extraneous shocks, the very fine and brittle balance pivot would break if not for the shock protection spring flexing under the pressure shock of the pivot, absorbing the force of the impact and then reverting to prior form, protecting the delicate balance pivot and associated parts.

THE ESCAPEMENT

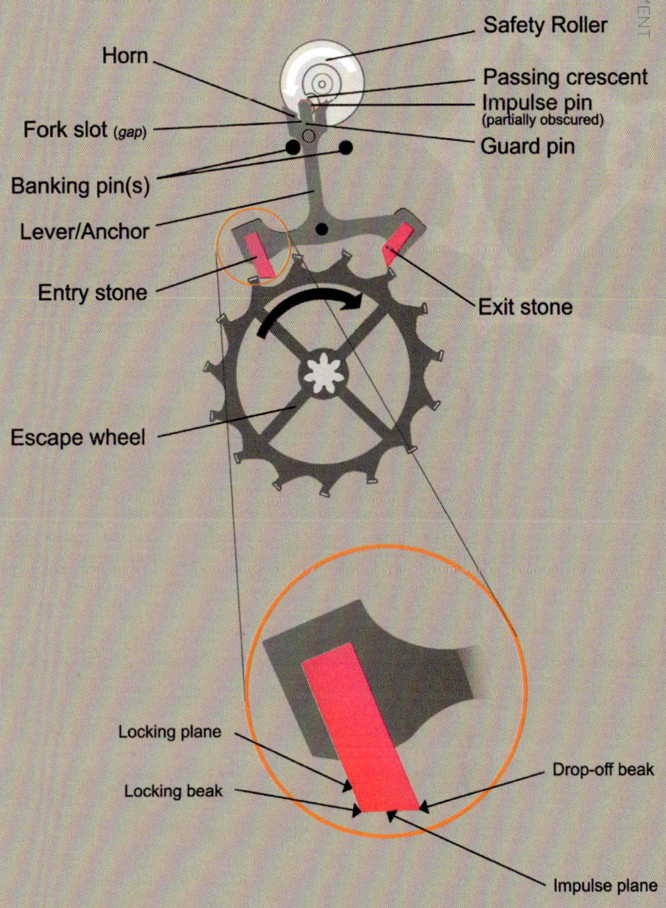

Safety Roller
Horn
Passing crescent
Impulse pin
(partially obscured)
Fork slot (gap)
Guard pin
Banking pin(s)
Lever/Anchor
Entry stone
Exit stone
Escape wheel
Locking plane
Drop-off beak
Locking beak
Impulse plane

SHOCK PROTECTION SYSTEM

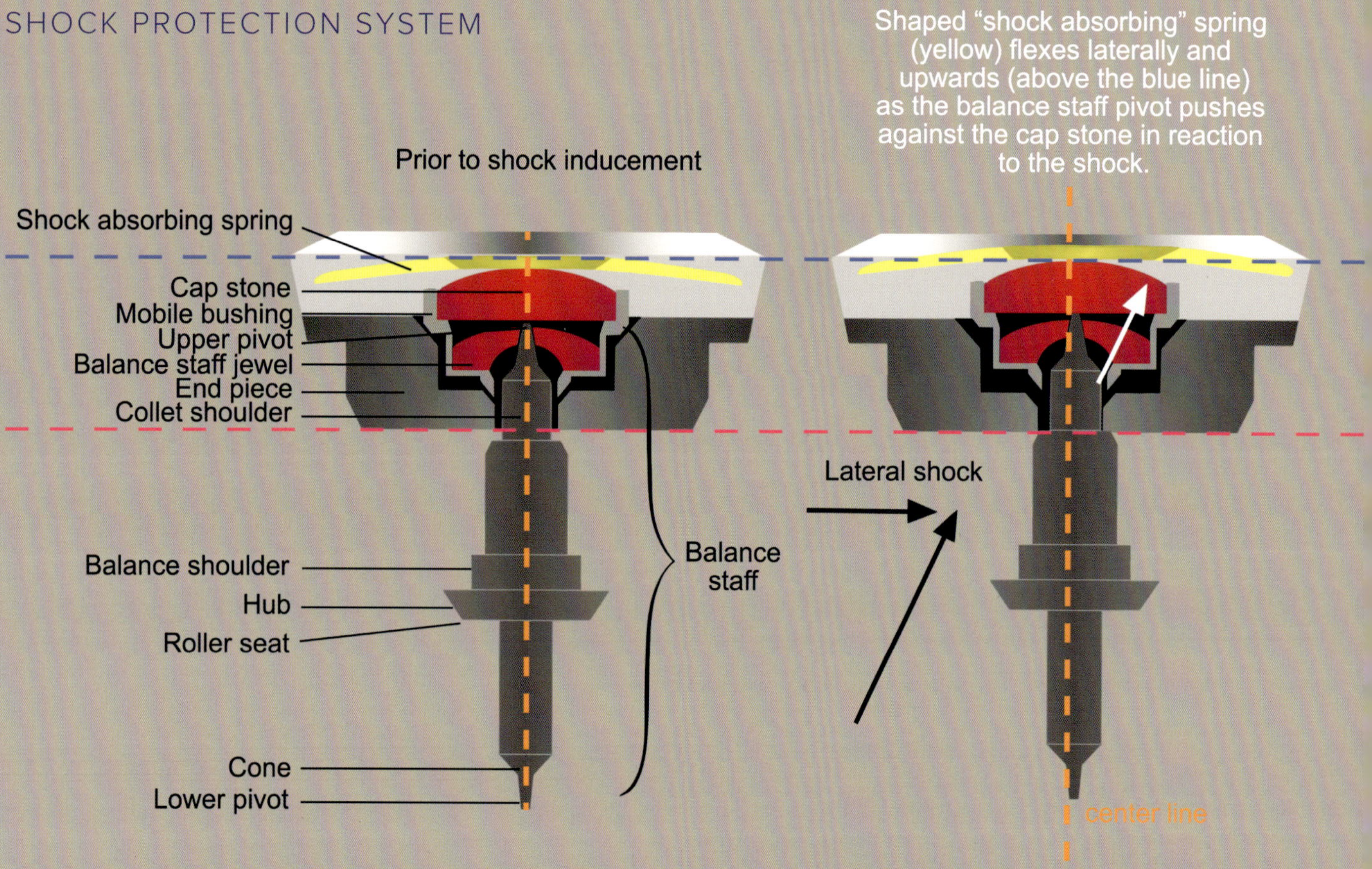

Shaped "shock absorbing" spring (yellow) flexes laterally and upwards (above the blue line) as the balance staff pivot pushes against the cap stone in reaction to the shock.

Prior to shock inducement

Shock absorbing spring
Cap stone
Mobile bushing
Upper pivot
Balance staff jewel
End piece
Collet shoulder
Balance shoulder
Hub
Roller seat
Cone
Lower pivot

Balance staff

Lateral shock

center line

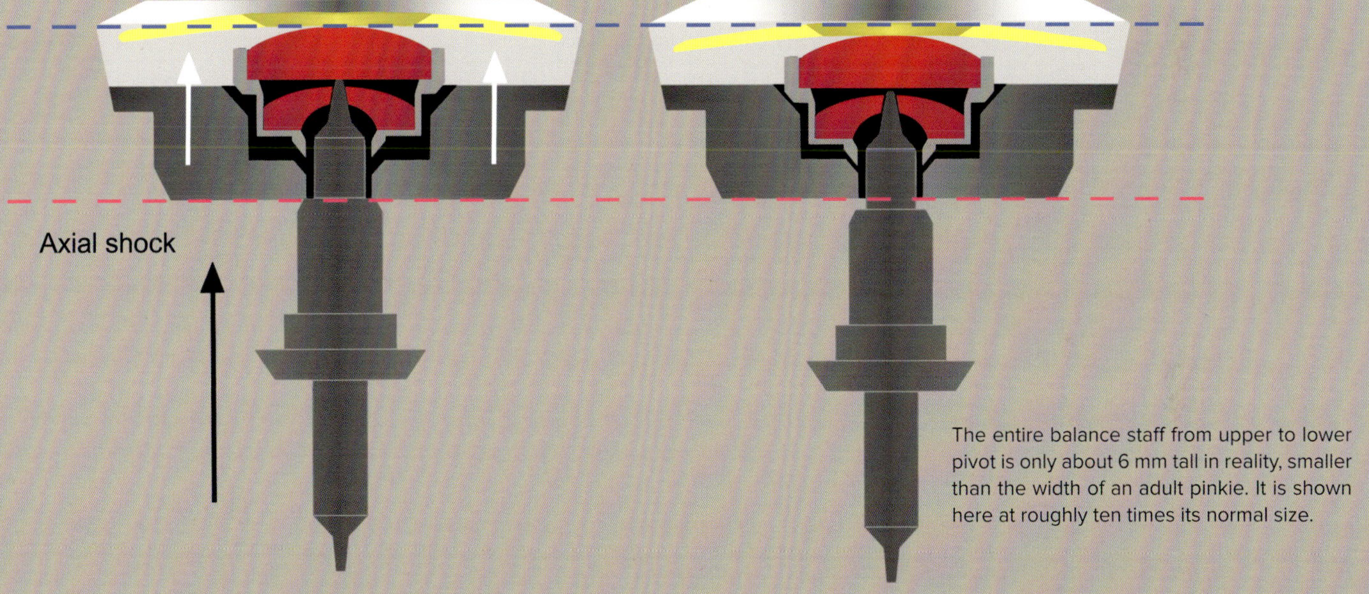

The entire balance staff from upper to lower pivot is only about 6 mm tall in reality, smaller than the width of an adult pinkie. It is shown here at roughly ten times its normal size.

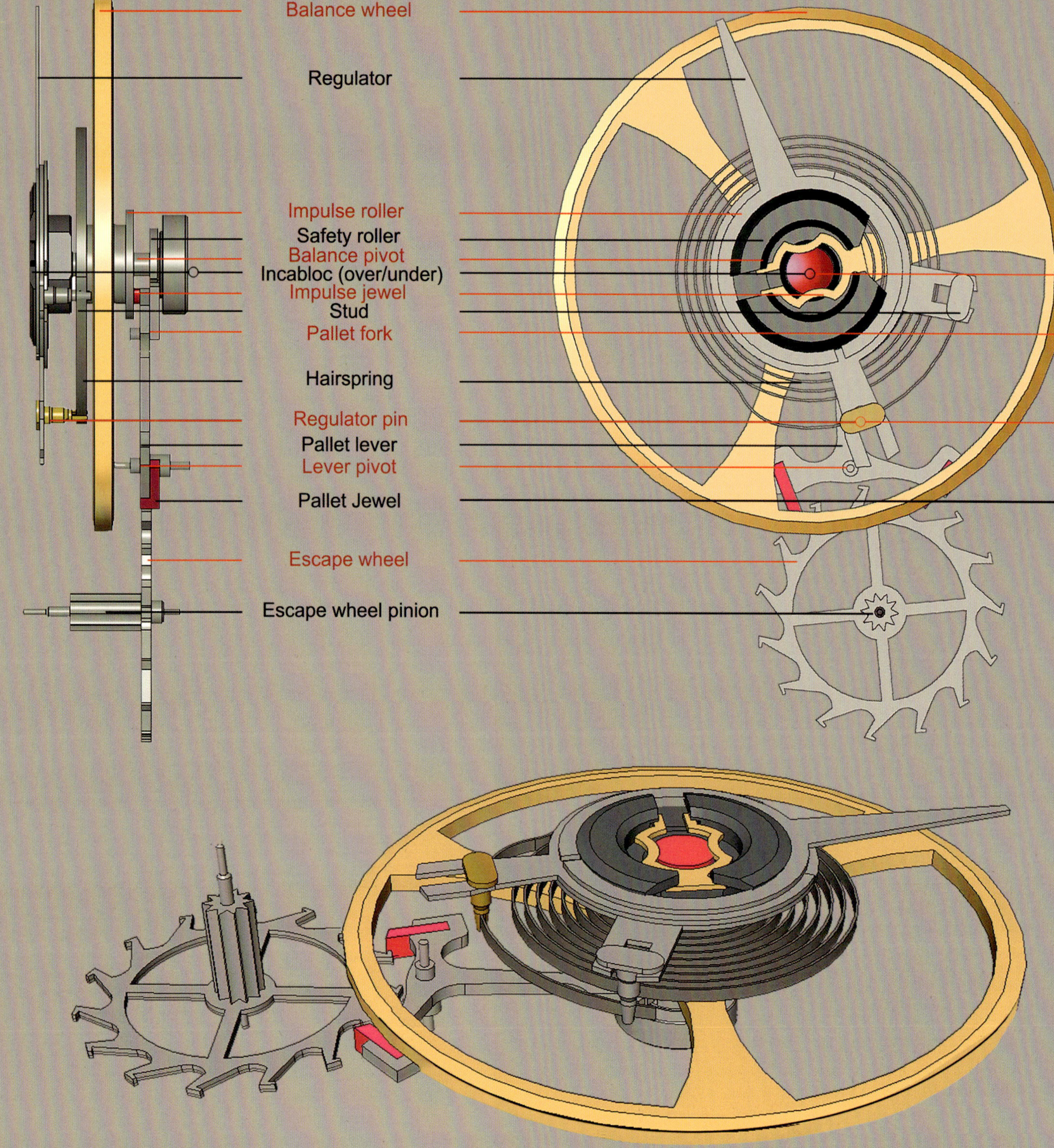
Balance wheel
Regulator
Impulse roller
Safety roller
Balance pivot
Incabloc (over/under)
Impulse jewel
Stud
Pallet fork
Hairspring
Regulator pin
Pallet lever
Lever pivot
Pallet Jewel
Escape wheel
Escape wheel pinion

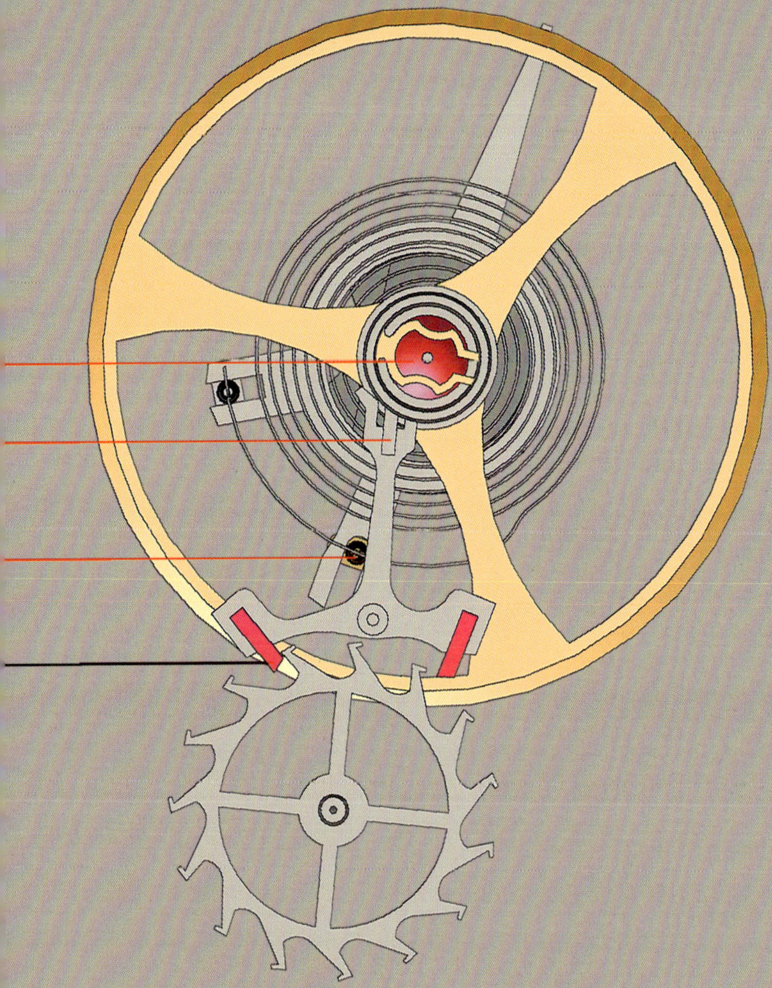

1 - The first sound is generated by the impulse pin (pink jewel), knocking against the fork slot (green), as part of the unlocking phase.

2 - As the escape wheel rotates, the escape tooth (blue) unlocks and slides along the locking plane (first recoiling slightly backwards), then falls onto the impulse plane (white), generating a second sound.

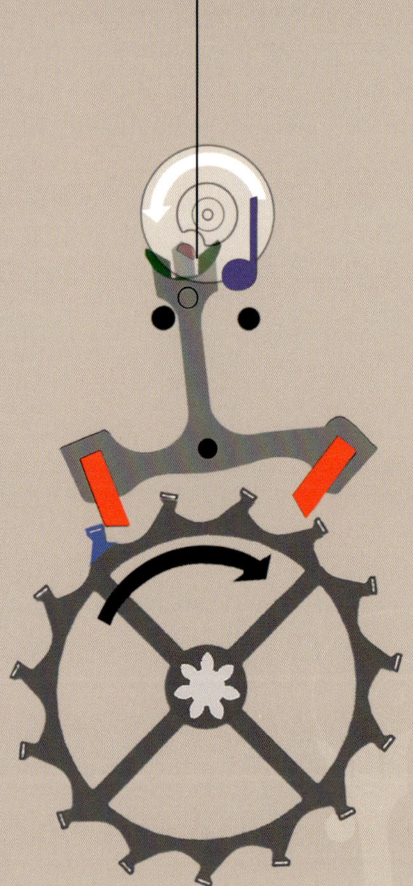

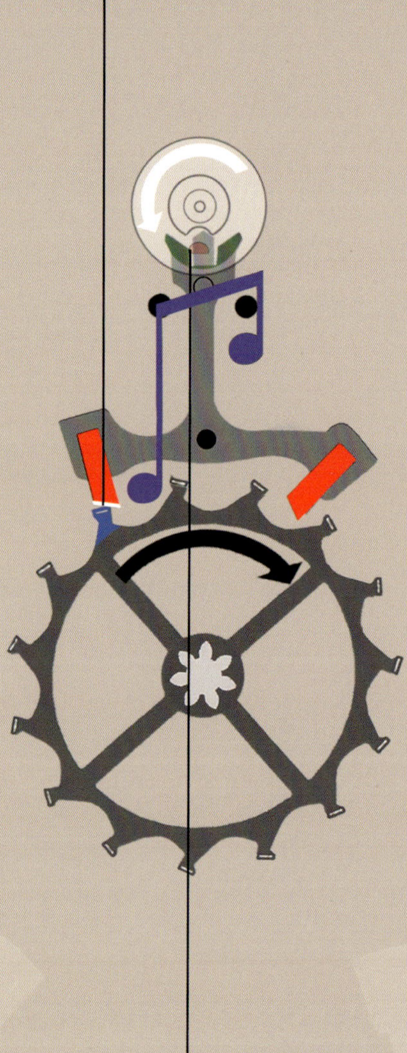

3 - A third "sound" is generated by the impulse pin hitting the fork slot and starting the impulse phase. To the human ear, this sound is indiscernible from the second sound, but it is detectible by a sensitive timing machine.

4 - Another tooth lands on the locking plane (white line) on the opposite stone (purple - 4th impact), generating a third discernable sound.

6 - The balance now oscillates in the reverse direction, rotating the roller and knocking the pallet fork to the opposite banking pin, starting the process anew.

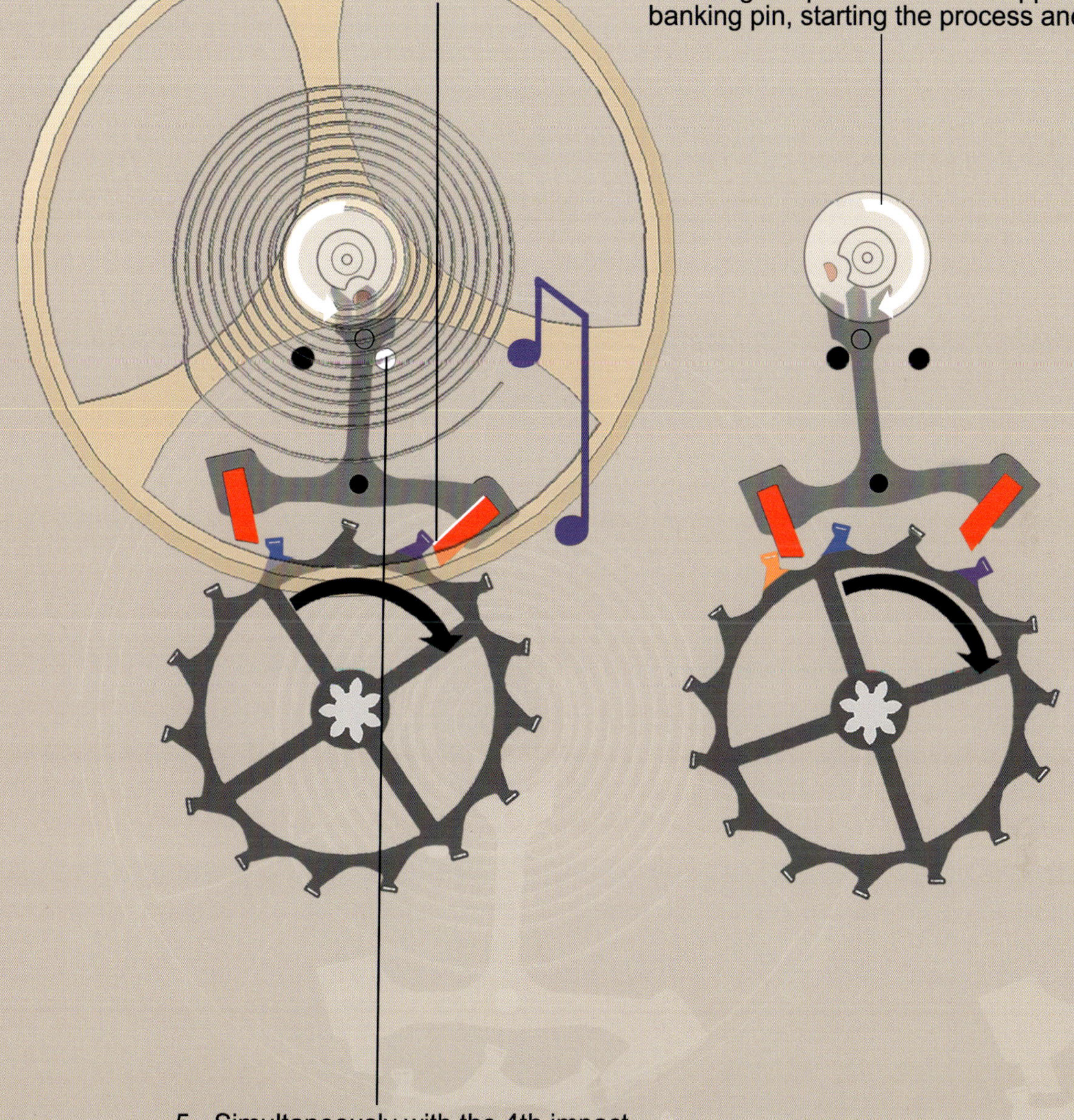

5 - Simultaneously with the 4th impact, the 5th impact of the pallet fork hitting the banking pin (white circle), amplifies the third sound.

The double-page image of the entire balance assembly illustrates top, side, and isometric views in order to show the major parts from common angles. The regulator index, shown with its long, pointy arrow, increases or decreases the effective length of the hairspring as it regulates the reciprocating rotation of the balance wheel. If the watch is running fast (gaining), then the regulator index is moved to slow it down (in the direction of the "-" marking, often engraved on the movement plates or elsewhere). Regulating a watch requires patience, practice, and (usually) only the tiniest of adjustments.

When we're taught about time as young students, our teachers usually tell us about the "ticktock" of the watch. In actuality, the ticking is really three discrete "tick" sounds repeating ad infinitum, more akin to tick-tick-tick than ticktock. If you ever listen to a mechanical watch (by placing the case back to your ear, or via a digital/electronic time-graphing apparatus), you will hear the very distinctive tick-tick-tick. Depending on the frequency of the movement, the ticking may be more or less rapid. It is fun to compare the sounds of a movement with 18,000 vph frequency (slower ticking intervals) to a movement with 36,000 vph frequency (faster ticking).

Last, the illustrated gear train (at right) is for the mathematically inclined. How exactly does a 2.5 Hz balance frequency translate into the motion of the second, minute, and hour hands? This last illustration should satisfactorily answer these questions for you.

While the operation of a watch movement may remain confusing to some, I do hope that these illustrations facilitate in elucidating the precise and complex interactions involved in the movement's functioning.

The balance (in this example) vibrates at 18,000 alternations per hour. Thus, it oscillates at 2.5 Hz, which translates to five back-and-forth movements of the pallet fork and escape wheel teeth locking and unlocking each second (18,000/60/60). Thus, the escape wheel's fifteen teeth require six seconds per single revolution. The escape wheel's pinion (in this example) has seven leaves (pinion "teeth"). Since the escape wheel pinion directly connects to the axle of the escape wheel, it correspondingly turns once every six seconds too. Thus, every minute the escape wheel's pinion A turns the second's hand (a) one revolution (70/7 × 6 = 60)—which makes intuitive sense—the seconds hand should rotate once per minute.

B, the pinion of the fourth wheel P, turns once per minute (as just explained above). Thus, every 7.5 minutes, Q turns once (60/8). C, the pinion of third wheel Q, therefore also turns once every 7.5 minutes. Thus, every hour, R turns once (80/10 × 60/8). This is represented in the motion of the minute hand (b). In order to move the hour hand once around the dial, the cannon pinion D mounted on the center wheel R must engage with the minute wheel Y, whose pinion E then engages with the hour wheel X, which revolves the hour hand (c) once every twelve hours (36/12 × 40/10 = 12).

Numbers in red are tooth count of wheels. Numbers in black are leaf counts of pinions. Arrows indicate direction of motion.

GEAR TRAIN ILLUSTRATED

(Dial Side)

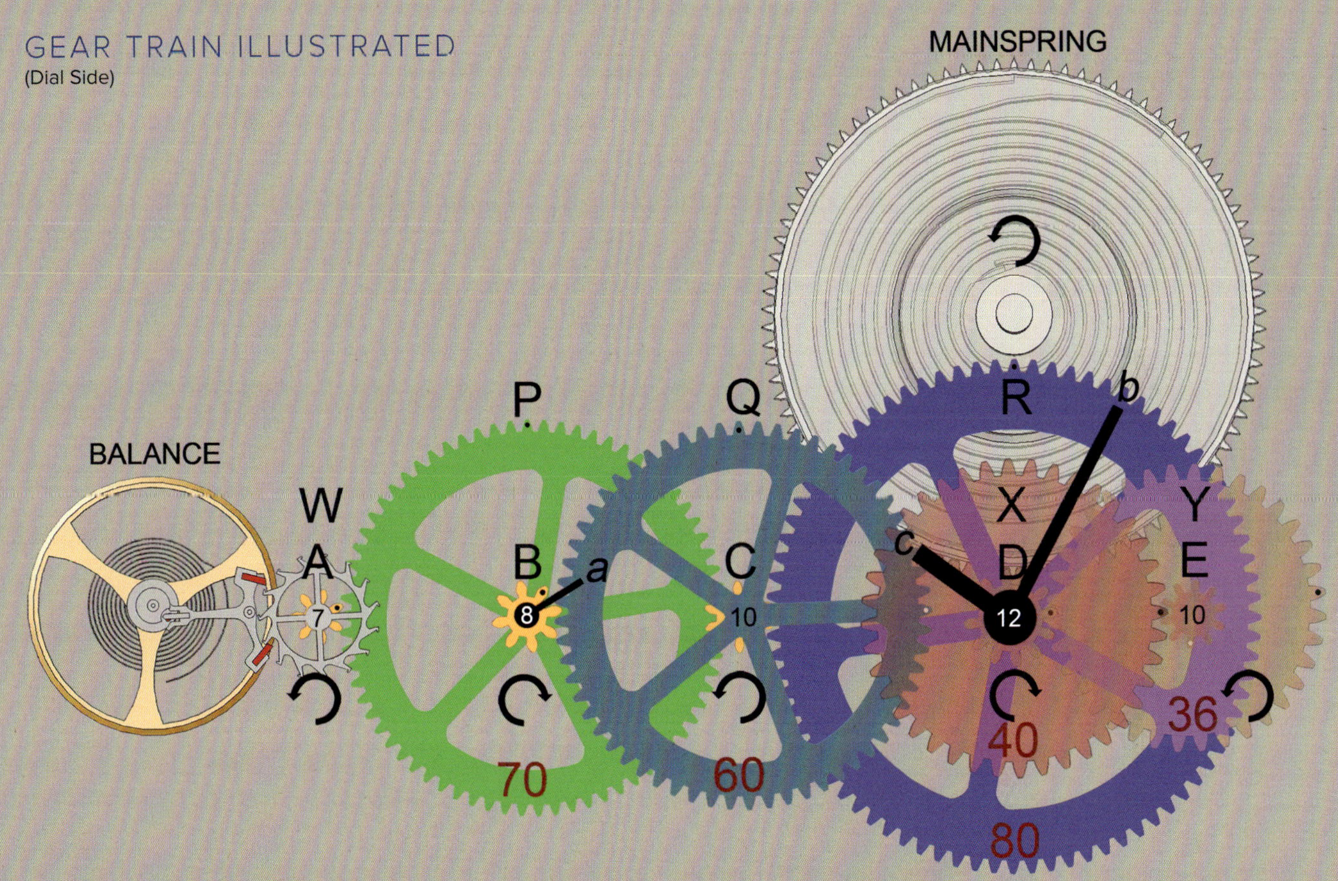

TYPES OF WATCH MOVEMENTS

QUARTZ

Quartz watches utilize an electronic oscillator regulated by a quartz crystal to mete out the time. A battery usually powers these watches. Quartz is the chemical compound silicon dioxide, prized for its relative immunity to temperature variation *and its piezoelectric*[161] *and amplification properties*. The invention of quartz timekeeping dates back to 1921. The first quartz clock debuted in 1927, and the first quartz wristwatch was unveiled in 1967. The first commercially available quartz watch was the Seiko Astron, released in 1969.

In most quartz watches, the frequency is precisely calibrated to 32,768 Hz (Hertz, or vibrations per second). The crystal is cut in the shape of a small tuning fork, and 32,768 is the result of 2 raised to the fifteenth power, or 2^{15}. A power of 2 is chosen so a simple chain of digital divide-by-2 stages can derive the 1 Hz signal needed to drive the watch's second hand. Driven by the frequency, a 15-bit binary digital counter will overflow once per second, creating a digital pulse once per second.

A typical quartz clock or wristwatch will gain or lose fifteen seconds per thirty days, or about half a second per day within a normal temperature range.

MECHANICAL (HAND-WINDING)

Mechanical watches, unlike their electronic quartz counterparts, utilize a mechanical mechanism with scores or hundreds of parts to measure time. This mechanism must be periodically wound to ensure that its mainspring (a tightly wound spiral ribbon inside a cylindrical barrel) possesses sufficient force to power the balance wheel through a series of gears. The balance wheel, which includes a fine spiral spring at its center—known as the balance spring or hairspring—swings back and forth, meting out the time in precisely controlled beats.

The magnificent movement housed within the Roger Dubuis Excalibur 45 mm Limited Edition Ref. EX45.79.9.9.71R in stainless steel (ca. 2008, originally priced at $83,850). Limited to eighty-eight numbered pieces, it features a complex mechanical movement with thirty-eight jewels, split-second chronograph, and automatic winding via a discreet microrotor. Looking toward the bottom right of the blue screw near the top of the image, one can just make out the engraved "Roger Dubuis" microrotor hiding within the movement.

161. The ability of a material to generate an electric charge in response to an applied mechanical stress. Derived from the Greek word *piezein*—to squeeze or press.

Typical mechanical watches oscillate at five, six, eight, or ten beats per second, which translates into 2.5, 3, 4, and 5 Hz, respectively, or, more commonly, 18,000, 21,600, 28,800, or 36,000 vibrations per hour.[162]

Prior to the introduction of marine chronometers in 1760 by the historically significant John Harrison, watches were not particularly resilient or reliable. However, toward the end of the nineteenth century, largely through the industrialization of the watchmaking process in the US, watches became much more reliable and accurate. Today's most accurate wristwatches are known as chronometers and are usually certified by one or more standard-setting bodies. These timepieces are accurate to within approximately five seconds per day. That may sound like a lot, but given that these timepieces are entirely mechanical and are based on an 86,400-second day, they are 99.994 percent accurate.

Calibre 1141 of the hand-wound Vacheron Constantin Patrimony Traditionnelle Chronograph in pink gold (Ref. 47192/000R-9352, ca. 2013, MSRP: $54,000). Manual-winding timepieces benefit from an unobscured view of the movement. The downside is that they require attentive regular winding, unlike their automatic counterparts.

Chronograph movements are particularly beautiful to behold because of the added mechanical complexity. The column wheel (making for a crisp start/stop of the chronograph) is visible with its eight triangular-shaped teeth at the twelve o'clock position.

162. The list includes only common oscillation frequencies; in 2011 and 2012, TAG Heuer produced two mechanical wristwatches with astonishing hourly vibrations of 3,600,000 and 7,200,000, respectively.

VACHERON
CONSTANTIN
GENEVE
SWISS

AUTOMATIC (SELF-WINDING)

The movement of one's wrist winds an automatic movement while worn. Also referred to as self-winding, watches with automatic movements utilize kinetic energy captured by the motion of one's arm and body to provide energy to an oscillating rotor to keep the watch ticking. Like their mechanical hand-wound counterparts, they are more satisfying (than quartz) to watch collectors because of the engineering artistry that goes into the hundreds of parts that make up the movement. With a fully wound mainspring, most automatic wristwatches can typically store about forty-eight to seventy-two hours of power.

Chopard Mille Miglia GT XL (Ref. 16/8997-3001, ca. 2008, MSRP: $5,650) with automatic winding.

The striped rotor affixed to the center of the movement and inscribed "Chopard" keeps the watch constantly wound via the movements of its wearer. Unfortunately, solid rotors obscure much of the beauty of the movement.

(5) POSITIONS
CHOPARD
MILLE MIGLIA GT XL
CERTIFIED CHRONOMETER
8997
330ft

TAG Heuer Monaco V4 (Ref. WAW2170.FC6261, ca. 2009, MSRP: 100,000 CHF). Limited to 150 pieces in platinum.

Although the movement is mechanical, the transmission system uses a novel belt mechanism designed to mimic the belt-drive system of an automobile engine.

The watch receives power from an oscillating rectangular ingot instead of a rotating rotor. The sliding mass winds a four-barrel system arranged in a V; hence the name V4.

TAG HEUER
V4
SWISS MADE
TAGHeuer

Ventura's Sparc MGS (microgenerating system), Ref. W51S (ca. 2011, MSRP: $2,600), combines the best of mechanical power with a modern liquid crystal display.

The inertia of the mechanical rotor's oscillating mass transfers to a tensioned spring that releases its force to a generator 17,000 times per day. An accumulator stores the energy and supplies a regular energy flow to the display.

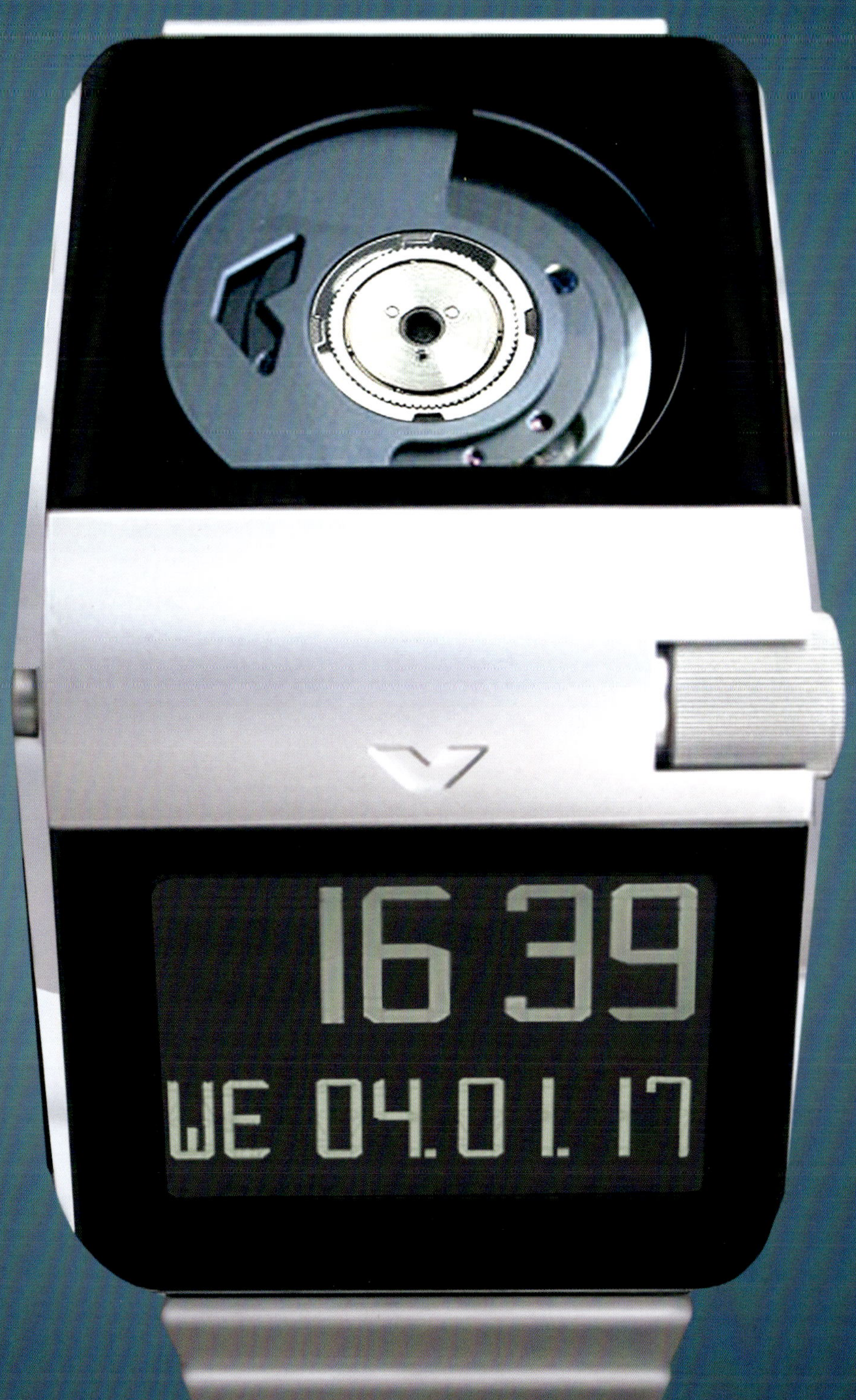
16 39
WE 04.01.17

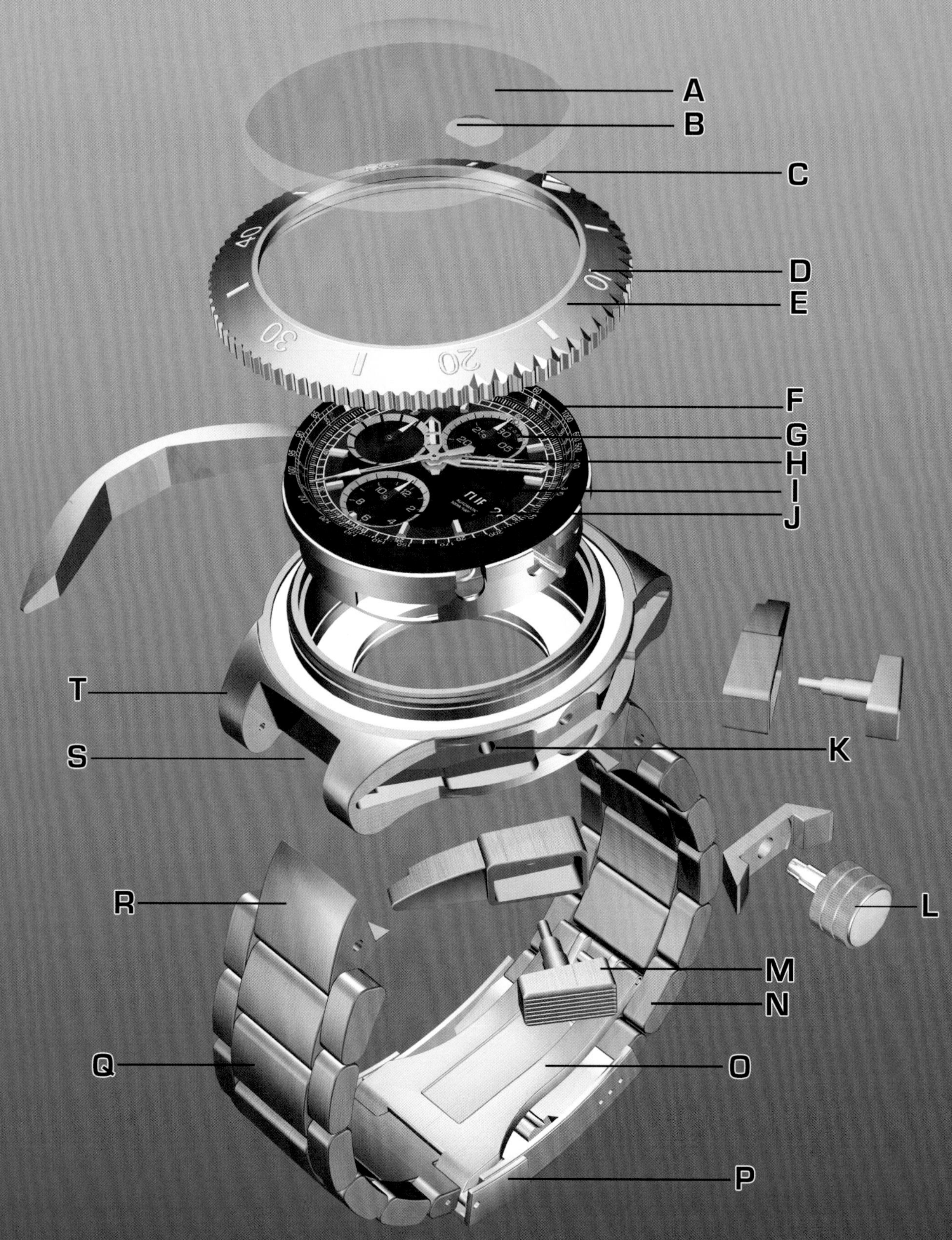
A
B
C
D
E
F
G
H
I
J
T
S
K
L
R
M
N
Q
O
P

PARTS OF A WRISTWATCH

A Crystal or Glass[163]

B Cyclops or Magnifier[164]

C Pip (indicates twelve o'clock on the bezel)[165]

D Bezel marker (luminous)

E Bezel (rotates bidirectionally or unidirectionally[166])

F Tachymeter—indicates velocity over a measured distance[167]

G Hour marker[168]

H Hands (there are usually three main hands—hours, minutes, and seconds)

I Day/Date aperture

J Subdial[169]

K Caseband (the side of the watchcase). The watchcase is the entire central structure that houses the movement (not pictured).

L Crown[170]

M Pusher (in this case, a chronograph reset pusher)

N Bracelet link[171]

O Deployant (or deployment[172]) clasp. This is a folding clasp.

P Safety clasp[173]

Q Bracelet (or strap in the case of nonmetallic materials)

R End-link (this link connects to the lugs)

S Lug width[174]

T Lug[175]

163. Usually synthetic sapphire (corundum) crystal in higher-end timepieces; mineral crystal in others.
164. Affixed to the crystal for date magnification; positioned above or below the main crystal.
165. May be any shape but is usually round or triangular, luminous, and raised above the bezel.
166. Rotates only counterclockwise in dive watches.
167. By comparing it to the position of the second hand and reading off the corresponding velocity. Also called a tachometer.
168. Usually with applied lume (vernacular for luminous material) in dive watches, but in dress watches, markers tend to be hand-applied without luminous material.
169. Subdials are usually lowered or raised above the main dial surface plane, but they may also be coplanar. They indicate additional functions such as power reserve or chronograph totals.
170. In dive watches, the crown tends to be of the screw-in variety to provide additional water tightness. The crown also features one or more gaskets (rubber rings) around the winding stem. Rolex famously introduced their Triplock crown in 1970 (debuting in the Sea Dweller; the Submariner followed in 1977). It features three O-ring gaskets: two positioned within the crown tube and one under the winding crown. The crown sets the time and other functions and usually has more than one horizontal position when pulled from the case (to differentiate the function being set).
171. Several bracelet links are usually removable to accommodate smaller wrist sizes.
172. The term "deployment" has become accepted parlance, but the word "deployant" is of French origin and is the original (many would say proper) way to refer to this type of buckle.
173. Almost exclusive to dive watches and added for additional protection in preventing accidental opening of the folding clasp while underwater.
174. The distance between the two lugs. It is a useful measurement when procuring a new or aftermarket strap or bracelet.
175. Usually rounded downward toward the wrist for a sleek look. Lug design affords watch manufacturers with an opportunity to be unique.

THE DESIGN PROCESS

"A common mistake that people make when trying to design something completely foolproof is to underestimate the ingenuity of complete fools."

–Douglas Adams, *Mostly Harmless*

Designers consider many things before embarking on a new project. They evaluate aesthetic, functional, ergonomic, environmental, economic, and even sociopolitical considerations.

The design process involves numerous activities, and one designer's methodology may differ significantly from another. Some designers prefer a collaborative design process, whereas others prefer to work alone. The dedicated solo designer would probably chuckle at the old proverb "A camel is a horse designed by committee" (an expression critical of group decision-making and referring to the camel's awkward[176] design).

Watch design is not particularly different from design in other disciplines. The same factors demand consideration when designing a watch, as when designing many other industrial items. The number of employees involved in the design process and the level of involvement of each individual will depend on the size of the organization and the importance of the project.

First is the preproduction phase. The product manager[177] briefs the designer, defining the intended goals and audience for the watch. The designer analyzes these goals and then researches and accumulates similar designs in the watchmaking field and in other fields; objects from nature as varied as leaves or insect wings or industrial objects such as windows, wheels, or pens may provide design inspiration. Over the decades, contemporary architecture and sculpture have also provided significant inspiration to watch designers.

It is important for the designer to maintain the brand's existing stylistic cues and identity. Today's Rolex Daytona, for example, is undoubtedly the stylistic successor to its antecedent (reference 6239) from 1963. The distinctive tachymeter with "UNITS PER HOUR" on the bezel, the tri-compax[178] subdial layout, the stick hands with luminous infill, the seconds track on the flange, and the oyster bracelet are all carried over (albeit with a modern interpretation) to contemporary Rolex Daytonas.

Ideas are brainstormed and analyzed, preliminary sketches are prepared, and the requirements are routinely consulted to ensure that the final design meets with the original objectives. Several solutions are modeled and presented to the product manager and related team members. Engineers are usually involved early on to see if the design is feasible, given watch movement size and layout, operability, and power constraints.

The design iterates[179] until the product matches its intended aspirations. When a design meets with the product manager's approval, it is often modeled in 3-D to see what the final watch will look like from every angle, in different lighting, and with different materials and color palettes. Three-dimensional modeling is not always

176. Albeit highly functional.

177. Product managers are trained professionals responsible for the development of an organization's product. Product managers own the business strategy behind a product, specify its functional requirements, and generally manage the project's feature set.

178. Interestingly, tri-compax (as in the Universal Geneve Tri-Compax from 1944) initially referred to something other than the arrangement of subdials: the triple (tri) complications of moon phase, calendar, and chronograph (within a four-subdial arrangement including running seconds).

179. Iteration is the repetition of the design process in order to produce the best possible outcome. It is a process of constant improvement and refinement.

performed, and its industry usage differs vastly from one organization to another. Numerous technical drawings of every part are then drawn in CAD.[180] It is usually at this stage that pricing constraints are deliberated and a bill of materials with costing is prepared.

The technical drawings are referenced in creating prototypes for the parts or to generate technical files for CAM.[181] Twenty-first-century manufacturing techniques usually skip the use of any kind of molds until the product is ready for mass production. Molds (which are used to produce hundreds or even thousands of identical parts between refurbishment) are very expensive to produce and are not usually utilized in prototyping. Typically, multiaxis machining[182] will sculpt the necessary prototype parts. These parts are then assembled into either a working or nonworking (far less costly) prototype, depending on the complexity of the project and the degree of preproduction certainty required. The prototype, a full-scale model of the watch, is evaluated in detail for overall fit and finish, use of colors and materials, size and thickness, durability and comfort, and ease of use and ergonomics.

Rolex probably has one of the most sophisticated prototyping departments in the watch industry. They have a team of at least two dozen planners, watchmakers, engineers, and modelers dedicated to producing full-scale prototypes, milled from brass slabs, and accurate to within 1/100 of a millimeter when compared with the design department's drawings. Even bracelets can go through twenty or more prototype iterations for a particular project.

The prototype may then be finalized for production, iterated further to accommodate additional requirements (or for further refinement), or shown to early customers to gather market research.

Production usually takes place immediately after the prototype's approval and sign-off. In order to minimize production costs, there is usually heavy utilization of third-party suppliers dedicated to the manufacturing of dials, cases, bezels, springs, jewels, gaskets, crowns, pushers, stems, screws, bracelets, clasps, crystals, markers, hands, and movements. Larger brands and watch groups produce almost everything in house, including movements. To the best of my knowledge, Seiko is the most vertically integrated watch company in the world (even surpassing Rolex's extensive in-house expertise). Seiko even produces the oils used for lubrication of its movement parts.

The design process from preproduction to prototyping may last anywhere from a few months to several years, depending on the number of iterations, the overall design complexity, and the project's goals. Some prototyping is merely conceptual, nothing more than a design study to explore directions for future models.

GOOD DESIGN

In the 1970s, the now-illustrious German industrial designer and university professor Dieter Rams pondered the question as to whether or not his designs were "good designs." Stemming from this question, he distilled his deliberations into ten recognized design principles.

Rams had successfully designed shelving systems, tables, lounge chairs, record players, and projectors (among a long list of many other products) for design firms Vitsœ (pronounced "vit-soo") and Braun. Rams has received many industry accolades and awards and has had his work exhibited at the Milan Triennale, the World Fair in Brussels, New York's MOMA, and San Francisco's SFMOMA.

"GOOD DESIGN" PRINCIPLES

Good design . . .

1. is innovative—*design evolves in tandem with improving technology*
2. makes a product useful—*good design is functional and should emphasize a product's usefulness*
3. is aesthetic—*beautiful objects are those that are well executed*
4. makes a product understandable—*good design is intuitive*
5. is unobtrusive—*a good design is a restrained design, leaving additional interpretation to the observer*

180. CAD or computer-aided design is the use of computers to aid in the creation, modification, analysis, or optimization of a design. CAD software increases the productivity of the designer, improves the quality of the design, advances communications through documentation, and creates a database for manufacturing.
181. Computer-aided manufacturing, also known as computer-aided modeling or computer-aided machining, is the use of software to control machine tools in the manufacturing process.
182. Multiaxis machining is a manufacturing process that involves tools that move in four or more directions and are used to manufacture parts out of metal ingots or other materials by milling away excess material with either a sharpened cutting head (the old fashioned way), water-jet blade, or laser (cutting)—or a combination of these techniques—among many other alternative forms of modern machining.

6. is honest—*it delivers on its promises*
7. is long lasting—*durability and timelessness can be synonymous, while avoiding fleeting fashions*
8. is thorough down to the last detail—*every detail is to be considered with care and accuracy*
9. is environmentally friendly—*the design of a product should minimize its environmental impact both physically and visually*
10. is as little design as possible—*minimalistic to the point of being as simple as possible, but no more*[183]

A wristwatch is a functional object that sees more user interaction than almost any other product. It is thus imperative that good design—and indeed these ten good-design principles—form an integral part of its creation. It is no accident that the most-celebrated watch designs embody the majority, if not all, of these principles.

THE GOLDEN RATIO

A discussion on design and aesthetics would probably be incomplete without at least an introductory discussion of the golden ratio.

Mathematically, the golden ratio (also called the divine proportion, golden mean, golden section, or golden proportion) is the ratio of the sum of two quantities relative to the larger of the two.[184] You will please forgive the mathematical description that follows, but I believe it will be worth the effort.

The Fibonacci numbers form a sequence (called the Fibonacci sequence), such that each number is the sum of the two preceding numbers (e.g., 0, 1, 1, 2, 3, 5, 8, 13, 21 . . .; i.e., 21 is 13 + 8 and 13 is 8 + 5). Eventually, the sequence will allow two consecutive numbers to provide a quotient of 1.618 . . .[185]

The ratio is usually shown as "a + b" is to "a," as "a" is to "b."

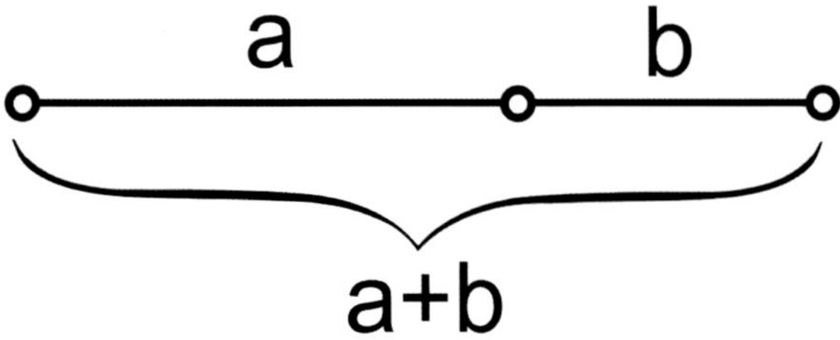

The previous description may be too abstract for many readers, so here is a crude, oversimplified example. Imagine a line that is 1,618 (millimeters or inches or any unit you choose—we will use millimeters, shown as mm). Now imagine that the line divides into segments such that "a" is 1,000 mm and "b" is 618 mm. Adding a + b, we get the sum of 1,618 mm, divided by "a" (1,000 mm) = 1.618. Further, a/b (a divided by b) = 1,000 mm / 618 mm = 1.618.

This ratio, the golden ratio, is an irrational number; when represented digitally—that is, with numerals in the form of a real number—it does not terminate (1.61803398875 . . .).

Where do we see this number come into play? Everywhere.

- The arrangement of seed spirals of a sunflower and seed pods on a pine cone
- The way branches grow from a central tree trunk
- The logarithmic spiral of snail and nautilus shells
- The spiral arms of the Milky Way galaxy (and other spiral galaxies)
- Hurricanes
- Human faces
- Fingers and their joints
- Even DNA molecules measuring 34 angstroms by 21 angstroms (for each double-helix spiral) present a ratio of 1.619.[186]

183. This last principle reminds me of Lotus founder Colin Chapman's most famous quote: "Simplify, then add lightness," a principle the famed car manufacturer still embodies today.
184. Represented by the Greek letter phi (pronounced "fi"; rhymes with pi—some say "fee"), shown in uppercase Φ (usually symbolizing the reciprocal) or lowercase φ or ϕ. Derived through a numerical series or mathematical/geometrical calculations.
185. Fibonacci numbers are named for the Italian mathematician Leonardo of Pisa (ca. 1170–ca. 1240–50), a highly regarded Middle Ages mathematician. He is also credited with introducing Arabic numerals to Europe. In his book *Liber Abaci* (1202), he utilizes this numerical sequence to model the population growth of rabbits on the basis of idealized assumptions.
186. Admittedly not exactly 1.618 . . .

Talal Ghannam, in his book,[187] augments and summarizes:

> *So the golden section is not restricted to the energy shells of individual hydrogen atoms, nor to the distances at which the electrons orbit the nuclei, nor even to the masses of its tiny constituents, it can also be found in the relative vibration frequencies between different adjacent atoms.*
>
> *From the shapes of animals and plants, to their growth rates. From inside the infinitesimal atoms and their constituents, to their configurations and vibrations. From the tiniest swirls of dust to gigantic galaxies. From the microcosm to the macrocosm, the golden section rules it all.*

That was a long but necessary preamble before we look at the golden ratio in design.

There is some disagreement regarding the modern usage of the golden ratio in design. Even the archetype, Da Vinci's Vitruvian Man, exhibits a ratio closer to the range of 0.606 to 0.609, smaller than the golden ratio's actual reciprocal of 0.618.

While applying the ratio diligently may lead to a holistic design, it is not necessarily always the case, or even often the case. Furthermore, beautiful and harmonious design is certainly possible without resorting to its implementation.

Notwithstanding this criticism, there are numerous examples (purposeful or not) of the golden ratio in design—including watch design.

Similar to the rule of thirds[188] in photography, once you familiarize yourself with the elegance of the golden ratio, as implemented in specific watch designs, it becomes difficult to unsee.

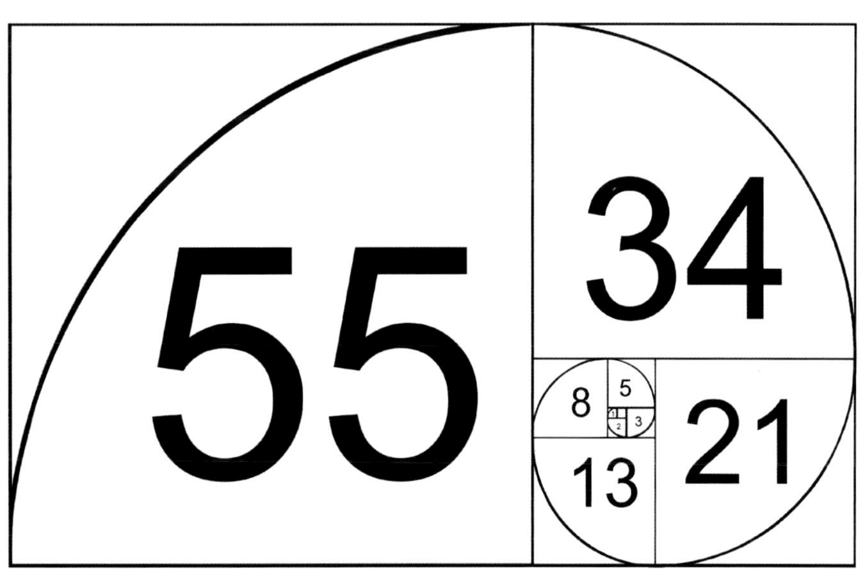

The golden ratio represented as a golden spiral (nautilus pattern) within a golden rectangle. Each higher-value square is equivalent to the sum of the previous two lower squares. As the numbers increase, the larger number divided by its adjacent smaller number will tend to the limit, 1.618 . . .

187. *The Mystery of Numbers: Revealed through Their Digital Root*, 2012.

188. The rule of thirds is one of the first techniques the amateur photographer learns. Although it is broken on many occasions, the rule does assist in creating harmonious shots, especially for novice photographers. The idea is to imagine a grid of "squares" three across by three down superimposed on the image. Placing points of interest in the intersections or along the lines of this imaginary grid, the photo becomes more balanced and provides a more natural interaction when viewed.

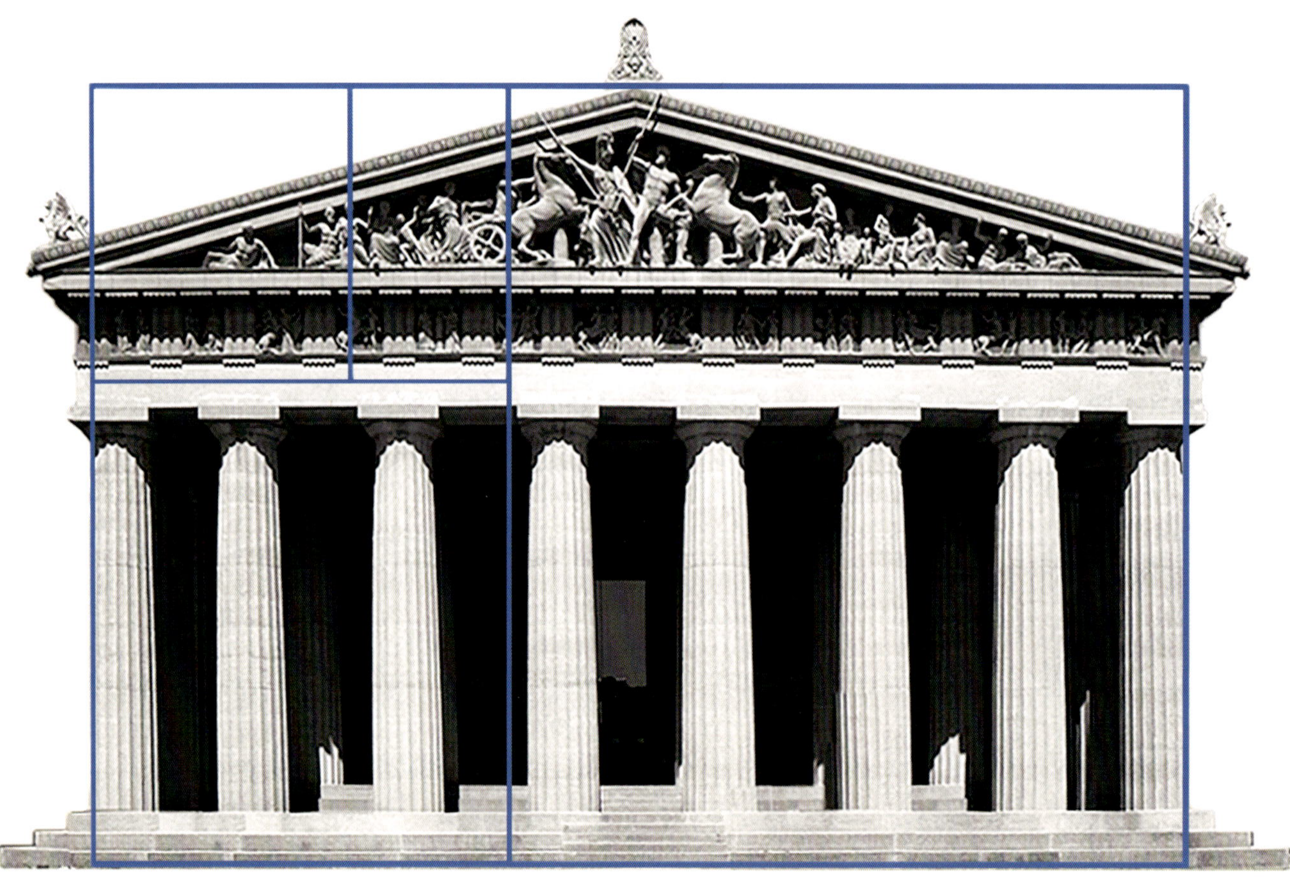

The Greek Parthenon, constructed ca. 447–438 BCE. There is some dispute regarding intentional consideration of the golden ratio in the building's planning and construction, both in the interior and exterior dimensions (*exterior only shown*). However, whether or not there was a deliberate intent to model the structure on the golden ratio, it is readily apparent that the building conforms (almost perfectly) to these proportions. Only when including the first and second steps does it lead to "perfect" proportioning.

Fun with geometry! A. Lange & Söhne's Grand Lange 1 utilizes the golden ratio (in the form of a golden triangle—the golden ratio appears in many geometric shapes) to create a harmonious design that "looks and feels right." Whether this design appeals to you or not, it is difficult to argue with the dial's overall coherence. Line segment "ab" divided by line segment "a" will result in the quotient φ. The "snowflake" design is the true center of the watch dial. It is intersected by the lines passing through x, y, and z (points of the red equilateral triangle circumscribed by the blue circle). Since this book is not intended as a mathematical treatise, I have not shown the geometric proof. See earlier discussion on phi.

F. P. Journe's Centigraphe utilizes the golden triangle in its dial layout. Observe that the pinions of the three subdials correspond to the inner triangle's intersection points.

The Jaeger-LeCoultre Reverso's subsecond dial and overall dial geometry implement aspects of the golden rectangle blueprint. It is no wonder that this straightforward yet harmonious design has been in production since 1931.

DESIGN ICONS

There is perhaps no twentieth-century watch designer more prolific or celebrated than Gerald Genta (1931–2011). Many credit him with creating the entire discipline of watch design.

His two most iconic and recognizable designs are, without doubt, the Patek Philippe Nautilus and the Audemars Piguet Royal Oak—penned by Genta several years apart and ushering in an era of the luxury sports watch.

Prior to the introduction of the Audemars Piguet Royal Oak, luxury watches were mostly fashioned in precious metal. As the quartz crisis of the 1970s threatened to upend the entire Swiss watch industry, brands struggled to differentiate.

The Royal Oak took Genta an evening to design.

The legend of the Nautilus is even more extraordinary. According to a retelling by Genta himself in a 2009 interview,[189] he created it for Patek Philippe executives in five minutes.

Genta also designed the Bvlgari-Bvlgari, Breguet's entire collection, Pasha by Cartier, the dollar watch by Corum, and watches for Chaumet, Van Cleef & Arpels, Hamilton, Bulova, and Timex.

The Patek and AP designs are aquatically inspired: the Royal Oak is reminiscent of a diver's helmet, and the Nautilus is reminiscent of a ship's porthole window—oddly similar. An analogous design for IWC's Ingenieur line followed shortly thereafter, incorporating some of the familiar design motifs from the Royal Oak and Nautilus. A depiction of all three timepieces appears below. These designs catapulted the wristwatch from functional time-telling tool to avant-garde artistic creation.

To illustrate the enormous impact of Genta's designs on the watchmaking industry, below the images of the three classic Genta creations is an image of the much more contemporary Hublot Big Bang, first released in 2005 with its bold 44 mm diameter case. In interviews, the acclaimed industry titan Jean Claude Biver has acknowledged that the Royal Oak and Nautilus served as inspiration in the creation of the Big Bang's daring design.

1972 Audemars Piguet Royal Oak. According to Genta, the inspiration came from a diving helmet viewport with screws on its bezel (*inset*) that he had seen prior to the AP design's creation.

189. "[The Nautilus] is a watch that I designed during the Basel Trade Fair. I was at the restaurant of a hotel and some people from Patek were sitting in one corner of the dining hall, while I was sitting alone in the other corner. I told the head waiter: 'Bring me a piece of paper and a pencil; I want to design something,' and I designed the Nautilus while observing the people from Patek eating! . . . It was a sketch that I completed in five minutes. . . . It very quickly met with success. I made the prototype in my studio, and its success was accelerated." http://www.veryimportantwatches.com/files/pdf/creating_desing_rules_en.pdf.

AUDEMARS PIGUET
AUTOMATIC
20
SWISS

Patek Philippe's Nautilus, designed in 1974 and presented in 1976. The inset image shows a round porthole (for reference purposes). Note the hinge to the left, mimicked in Patek's hinged case design.

Genta purportedly based Patek's "oval" porthole design on Jules Verne's *Nautilus* submarine, helmed by the fictional Captain Nemo. Verne appropriated the name "Nautilus" from the first practical submarine in history, designed by engineer Robert Fulton on commission by Napoleon Bonaparte (ca. 1800). Verne's description of the *Nautilus* in his book *20,000 Leagues under the Sea* was reportedly based on a model he saw of the newly developed French navy submarine *Plongeur*, which was presented at the 1867 Exposition Universelle.

The IWC Ingenieur SL (for Steel Line), introduced in 1976–77 in response to AP's and Patek's luxury steel watches. For the third generation of Ingenieur, Genta created a gorgeous checkerboard guilloche dial and five utilitarian recessed holes (for bezel removal) on the otherwise unadorned bezel—reminiscent of his earlier work for Patek and Audemars Piguet.

Under the tenure of Jean Claude Biver, in 2005 Hublot introduced the impactful Big Bang (Ref. 301.SB.131.RX, MSRP in 2020: $13,600, 44 mm case diameter), the spiritual successor to the original Royal Oak and Nautilus forty years prior. While Hublot had previously utilized a porthole design as its main motif—indeed, the word "Hublot" is French for porthole—the Big Bang was a giant leap forward for the brand, creating what has become its signature look and best-seller. It embraces the bigness trend of the early aughts and is an exemplary specimen of the period's design sensibilities.

The "fusion" of twenty-first-century materials such as Kevlar, carbon fiber, rubber, ceramic, and steel was a bold move, and one that was subsequently imitated by numerous brands.

HUBLOT
GENEVE
SWISS MADE

DESIGN AND COLOR TRENDS

Every decade has a core style with which it is associated. Some style trends endure for several decades; others have a short life span but are reborn many years later as a revival or reinterpretation. From a historical perch, it is easier to spot the central design cues of a particular era.

Design shapes and *is shaped by* all aspects of socioeconomics. Watchmaking is no exception, but given a longer product development cycle often measured in years (sometimes bordering on decades), it usually takes a while before contemporary fashion impacts Haute Horlogerie.[190]

Art nouveau (merging the fine arts with the decorative arts) influenced art, architecture, the decorative arts, and industrial design, including watchmaking, from the late 1890s through the early 1910s. Women's watches and, in particular, women's pocket watches adopted the often-flowery curvilinear style. At the start of the twentieth century, pocket watches remained the dominant means of portable timekeeping, despite the contemporaneous introduction of "trench watches." Trench watches began as modified pocket watches involving soldering wire lugs to watchcases to enable the attachment of a strap—primarily for military personnel who required easy access to a timekeeping device while on the battlefield.

Cartier's Tank Normale (ca. 1920). Prior to its introduction, the traditional method for attaching a strap to a watch was via a connection to soldered wire lugs on the case.

The Tank, modeled after the silhouette of a World War I French tank (as viewed from above) became an instant classic. Cartier manufactures a version of the Tank to this day. It is an undisputed icon.

190. "Haute Horlogerie" literally translated means "high watchmaking" (in French). It is a term often associated with luxury timepieces exhibiting multiple complications, avant-garde design (of a technical or decorative nature, or both), or meticulous artisanship. Sometimes a timepiece features all the aforementioned characteristics.

CARTIER

Watchcases and dials embraced art nouveau's decorative embellishments. Flowery motifs adorned numerous pocket watch bezels, and hunter-case designs grew considerably more elaborate and decorative, with engravings often ensconcing entire surfaces. The watches incorporated precious metals, precious stones (such as diamonds and sapphires) and semiprecious stones (such as lapis lazuli), gilt floral ornamentation, painted enamel (especially colorful and finely detailed portraiture and landscapes), and elaborately fashioned fobs.[191]

The 1920s birthed the art deco movement with its rectangular ornamented aesthetic—straight lines and corners as opposed to the flowing, decorative, and floral ornamentation of the outgoing style. The golds and yellows of the Roaring '20s festooned signs, posters, edifices, and clothing. Vibrant blues and reds beautified these designs. Watches embraced gold (gilt) dials and lettering on their geometric forms. Timepieces that stand out for their historical impact include Cartier's square Tank and Rolex's cushion-shaped (and water-resistant) Oyster.

Patek Philippe 18K yellow gold rectangular wristwatch ca. 1920. This style was produced in a variety of sizes between 1902 and 1930 for Gondolo & Labouriau, a prominent Brazilian retailer. Case measures 43 by 30 mm and is curved and hinged with thin wire lugs connecting to the strap.

The early 1910s witnessed the transition from "art nouveau" to "art deco," and this reference is an excellent exemplar of the period. The exploding art deco numerals surrounding the oval minute track experienced somewhat of a revival in the early twenty-first century, exemplified by timepieces from Franck Muller. Timepieces like these are quite difficult to acquire, since very few examples come up for auction.

191. A chain attached to a watch to enable carrying in a waistcoat or waistband pocket. Early examples often included a small ornament attached to the watch chain.

PATEK · PHILIPPE &
GENEVA-SWITZERLAN

The adverse economic impact of the Great Depression stifled the exuberance of art deco's ornamentation. Trying to stay positive during the despair of a difficult decade and with the popularization of Technicolor in movies such as 1937's *Snow White and the Seven Dwarfs* and 1939's *Gone with the Wind* and *The Wizard of Oz*, primary colors were still strong and pervasive, but practicality and simplicity supplanted decoration.

Rolex introduced the chronometer-certified Prince, with its elegant rectangular case; Jaeger-LeCoultre introduced its still-manufactured Reverso; and Patek Philippe offered the curved rectangular "Staybrite" steel watch.

In the 1940s, primary colors dominated war posters and comic books in an effort to stimulate patriotic optimism. The world economy, restrained by the effects of World War II and influenced by a production shift from civilian manufacturing, generated simple, stark watch designs—mostly in traditional round cases. Watch companies pivoted to manufacturing military watches and chronometers, fuses and sapphire bearings for artillery, and altimeters and other instruments for the aviation industry.

The year 2015 marked the seventy-fifth anniversary of the IWC Portugieser (German for Portuguese). The 1930s was a turbulent decade for the watch industry. Portuguese wholesalers, looking for wristwatches as precise as marine chronometers, approached IWC. That watch would become the 325 (although original case back innards were engraved "228").

These watches commenced delivery in February 1942. Ref. 325 with Calibre 74, shown here, is from ca. 1942. The Portuguese is an iconic IWC watch and, as late as the early 1990s, was still one of the biggest-diameter wristwatches at the time. When first introduced in the early 1940s, case sizes were generally 33 mm or less. The 43 mm Portuguese, which uses a pocket watch movement, was the precursor to modern oversized timepieces prevalent since the turn of the millennium.

International Watch Co
SCHAFFHAUSEN

According to Omega's records, the British Ministry of Defense received more than 110,000 timepieces from the Swiss brand during the war. The 1948 Seamaster was the civilian manifestation of war-tested combat timepieces and incorporated the lessons learned about water resistance, antimagnetic shielding, and reliability on the battlefield.

Examples from this period exhibit dials with or without the famous "Seamaster" script. For the first time, the watch featured a rubber gasket. Seamaster is Omega's first collection of watches still in serial production today. Until about 1957, the Omega Seamaster remained a dress watch.

With the introduction of their trilogy of the "professional" line of watches in 1957, Omega launched the Seamaster 300, the Railmaster, and the Speedmaster. Watch buyers frequently compare a modern professional Seamaster with Rolex's Submariner when considering a luxury dive watch acquisition. The flagship of the contemporaneous line, the "Omega Seamaster Planet Ocean," is the modern expression of many of the best features and capabilities of Omega Seamasters of the past.

OMEGA
AUTOMATIC
SWISS MADE

US watchmaking supremacy would not return from the manufacturing shift, and post–World War II, the Swiss secured industry dominance.

With the onset of the jet age, the 1950s brought international travel to the masses, accompanied by a renewed sense of global optimism. In the US, returning soldiers made use of the newly introduced GI Bill to help pay for college, graduate school, and training programs. Thousands of newly minted graduates bought homes, started families, and launched new businesses. Frugality diminished and memories of the lean war years began to fade as the positive economic effects brought about by a bourgeoning middle class accelerated production and wealth accumulation. The warmth of pastel shades ruled color preferences. Light pinks and salmons, powdery blues, and muted yellows adorned cars, appliances, and interiors. Patek Philippe's reference 2523 world timer and Omega's Speedmaster Broad Arrow are two well-known introductions from the 1950s.

At the request of the famous but now-defunct Pan Am, and to mitigate jet lag, Rolex teamed up with the airline to design a watch for their long-haul pilots. When traveling internationally, the new reference would allow pilots to remain on "home time" while still being able to read off local time. The Rolex 6542 GMT-Master (original MSRP: $240), which launched in 1954, is the result of their efforts. Its long red hour hand displays the additional time zone via the twenty-four-hour bezel markings.

This reference is the "original" GMT, and mid- to late 1950s models are especially sought after, since they feature a rare (and heat-sensitive) Bakelite (the first synthetic plastic) bezel—made even more atypical because it was recalled for purported radioactivity. The Bakelite bezel helped eliminate glare. Extant versions of these timepieces without "cracked" numerals in the Bakelite or without temperature-induced discoloration of the "Pepsi" bezel are exceedingly rare. In good condition and with original paperwork, 6542s fetch upward of $250,000 at auction.

Rolex subsequently discontinued the Bakelite bezel and replaced it with a more durable metal insert instead. Modern versions feature ceramic bezels.

ROLEX
GMT-MASTER
OFFICIALLY CERTIFIED
CHRONOMETER
SWISS

A travel watch from the same decade as Rolex's GMT-Master, Patek Philippe's Ref. 2523 was originally launched in 1953, supplanting the less complex single-crowned Ref. 1415. The 2523 ended production around 1966. The suffixless model designation and its "compact-lug" sibling, the 2523-1, are fashioned from a solid ingot of pure unobtanium. Not really, but they may as well be, so rare and collectible are they.

In rose gold, there exist only seven known examples of the 2523. A handful of additional specimens survive in yellow gold (eighteen), and only a single known timepiece exists in white gold. The 2523-1, shown here, exited the factory in equally anemic quantities. The 2523s appeared in a variety of enameled dials, whereas the 2523-1s incorporated silvered or gilt dials. Bracelets adorned a tiny fraction of these. In addition to the two dozen or so known examples, combined there are probably a dozen 2523s and 2523-1s remaining unknown to the collecting community, but not many more.

The 12 400 HU manual-wind movement is a marvel designed by celebrated watchmaker Louis Cottier and based on his "Heures Universelles" (world-time system—patented at least a decade prior).

The present-day popularity of this ultrarare watch is ironic. Upon initial release, the watch proved unpopular, and few examples ended up selling. However, as is often the case, one generation's ugly duckling becomes another generation's swan. The auction world loses its collective mind whenever a 2523 comes up for sale.

Setting and using the world-time complication is straightforward. Moving the crown located at the three o'clock position results in the twenty-four-hour central ring and watch hands moving around the dial in opposite directions. The other crown, located at the nine o'clock position, rotates until the local city appears on the city ring at twelve o'clock. The current time of each world city can now be accurately established. In the image shown, the local time in Samoa is 10:08 a.m. If you wish to know the current time in London, it is 9:08 p.m. (London sits almost 180 degrees opposite Samoa on the city ring).

The left crown together with the 35.5 mm case size—large for the time—were likely the main reasons for the watch's initial commercial ambivalence. If you happen to have one squirreled away in a vault somewhere, it is probably worth in the neighborhood of $3 million, assuming the most-recent auction sales are any guide. Ultrarare one-off examples have commanded much more: Christie's sold a unique Patek Philippe Gobbi Milan with blue enamel dial in Hong Kong in 2019 for around $9 million.

PATEK PHILIPPE
GENÈVE
SWISS
SAMOA ALASKA
TAHITI KLONDIKE CALIFORNIA
MEXICO DENVER CHICAGO
NEW-YORK MONTREAL BUENOS-AIRES RIO DE JANEIRO
AZORES MADEIRA DAKAR
ICELAND LONDON PARIS ALGIERS
OSLO GENEVA ROMA ISTANBUL CAPETOWN
BERLIN MOSCOW ADEN REUNION
MAURITIUS ISLD CEYLON
BOMBAY CALCUTTA SINGAPORE SAIGON
PEIPING TOKYO ADELAIDE
SYDNEY MARSHALL ISLD AUCKLAND ISLD FIJI ISLD

Design flair permeated the 1960s as new shapes and designs expressed the era's less restrictive social outlook and more inclusive role for women, especially as they engaged in the workforce in larger numbers. Interest in 1920s art deco saw a resurgence. Vivid psychedelic earth tones filled the color palette. Yellows, oranges, browns, and greens took center stage. Doxa introduced the SUB 300 at BaselWorld in 1967—purportedly the watch of choice for dive pioneer Jacques-Yves Cousteau, not least because of its superior legibility intensified by its effervescent "Doxa Orange" dial.

Form follows function. Doxa Sub 300T Professional (ca. 1968) with distinctive '70s tonneau case and "screaming orange" dial. In the mid-'60s, in consultation with Jacques-Yves Cousteau—renowned French naval officer, explorer, documentarian, researcher, and coinventor of the Aqua-Lung SCUBA apparatus—and diving legend Claude Wesly, Doxa embarked on creating a purpose-built dive watch.

The example shown here exhibits one of several "special" dials created for the US export market's burgeoning civilian dive community and Cousteau's own Calypso dive team. The "aqualung U.S. divers co." insignia on the dial was the trade name of Cousteau's US-based dive company. The Doxa-orange dial resulted from numerous tests performed in the murky waters of Lake Neuchatel, in which Doxa evaluated the visibility of various dial colors.

The oversized minute hand is one of many meticulously designed elements of this purpose-built timepiece.

The bracelet includes an imaginative flex buckle and a ratcheting mechanism for quick sizing without link removal. The "beads-of-rice" bracelet is a Doxa hallmark, carried over recently in the fiftieth-anniversary Sub 300 reissue. The original watch is water resistant to 300 meters, and the case is fashioned from a single ingot of stainless steel. The numbers on the bezel indicate the "no-decompression" times according to the US Navy Dive Table. Thus, at a depth of 120 feet, fifteen minutes is the dive time limit before necessary resurfacing.

DOXA
"aqua-lung"
U.S. divers Co.
SUB 300T
professional
FT

Rolex's Cosmograph Daytona and Heuer's Carrera with panda dials (silver dials with black subdials or black dials with white subdials) are two very collectible examples from the 1960s.

Rolex Daytona 6239 (ca. 1967). The quintessential 1960s Rolex collector's piece with "71" riveted bracelet and "Paul Newman" dial. Early-twenty-first-century collectors have pushed up the value of these chronographs almost tenfold in the last two decades. Expect to pay at least $100,000 for average examples.

The example illustrated here is mint with pristine tachymeter numerals, markers, and original lacquer; it would probably fetch well north of $300,000 at auction, and double that with exemplary provenance.

UNITS PER HOUR
ROLEX
COSMOGRAPH
DAYTONA
T SWISS T

Then came the 1970s, with the quartz crisis deeply rocking the wristwatch industry and with many brands disappearing or being forced to consolidate with competitors. I have already covered the designs of Patek Philippe's Nautilus and Audemars Piguet's Royal Oak, two of the most notable examples from this era, but there are many other interesting timepieces from the period: Breitling's Chronomat, Transocean, and Chronomatic; Heuer's Autavia and Monaco; Omega's Ploprof; Tudor's Oysterdate; Rolex's GMT Master; and Zenith's El Primero.

The vibrant colors that dominated the 1980s began taking hold in the mid- to late 1970s, typified in flamboyant style with bright polyester clothing, colorful logos (including Apple Computer's iconic "six-color" 1977 logo), and the polychromatic palettes of Lego, Hot Wheels, and Sesame Street.

Heuer's Monaco (Ref. 1133B), introduced in 1969, took '60s fashion—and an innovation allowing square cases to be waterproofed—and made a statement with a sharp-edged, curved-square, satin-brushed case.

The image shows the timepiece known by many as the "Steve McQueen." This is an example of the iconic watch the famous actor and racecar driver wore on the set of and in the movie Le Mans. The crown is intentionally placed on the left, since once worn, the automatic rotor takes care of winding.

MONACO
HEUER
10 12 2
8 6 4
25 30 5
20 15 10
AUTOMATIC
CHRONOGRAPH
SWISS
15

The '80s and '90s incorporated some stylistic cues from the 1970s but with a shift toward bright colors and geometric shapes and patterns. The Memphis Group, an Italian-based design collaborative active from about 1981 to 1987 and founded by Ettore Sottsass (a big proponent of radical design), introduced postmodernism to furniture, fabrics, lighting, ceramics, glass, and metal. While somewhat obscure to the general populace, the group's influence was historically impactful. Their furniture was often highly impractical for comfort (or safety), and only a single piece of their furniture ever made it to mass production: their "First Chair." The oft-derided '80s postmodernist look has been called a shotgun wedding between Bauhaus and Fisher-Price,[192] exemplified in traditional watchmaking by the eccentric and highly colorful watch designs of architect and designer Alain Silberstein, who utilizes primary colors and geometric shapes, especially in his dial hands, crowns, and function pushers. One could be excused for thinking that Swatch, Silberstein, and the Memphis Group collaborated. Swatch began dominating the analog watch industry with its colorful and inexpensive quartz timepieces.

Introduced to the world in 1989 (six years after the first Swatch watch), the plastic Swatch "Flumotions" GN102 is characteristic of Swatch's theatrical design and is not too dissimilar from the geometric and colorful stylings of the Memphis Group. The strap features a bright-yellow tang buckle with translucent strap keeper.

The timepiece consists of a scant fifty-one parts and ushered in an era of the fashionable and inexpensive Swiss quartz-analog watch. As of 2020, there are at least 4,200 extant references of Swatch. With initial prices from $30 to $40, fashionistas and collectors alike could stock up on several examples without breaking the bank. With the number of Swatch watches sold to date approaching one billion units, the brand continues to experience success.

192. Bertrand Pellegrin, "Collectors Give '80s Postmodernist Design 2nd Look," SFGATE, January 15, 2012.

© SWATCH AG 1988
swatch
SWISS

Casio's G-Shock and calculator watches also adorned many wrists, replacing less accurate analog mechanical timepieces with superaccurate and inexpensive digital quartz alternatives. The 1987 stock market crash probably marked the end of the postmodern era, as the recession brought about a decline in interest in the art market and the uber-consumerism of the late 1980s.

Rolex Oyster Perpetual Cosmograph Daytona (Ref. 16519, ca. 1997) 40 mm diameter, 12 mm thickness. A substantially upsized version of the Daytona since 1988's refresh—the case diameter was previously 37 mm. Featuring sapphire crystal glass, COSC-certified movement, tachymetric scale on the bezel, applied 18K white gold Arabic numerals and lozenge markers (at three, six, and nine o'clock), and 18K white gold case and deployant buckle.

The collector community currently underappreciates this reference, though interest is intensifying. The Zenith (El Primero)–powered Daytonas (Rolex Calibre 4030—detuned from 36,000 to 28,800 vph) from the late 1990s are very popular in steel and yellow gold, but the white gold version shown here receives far less admiration or interest.

To the novice, it looks like a standard steel Daytona. What makes this timepiece rarer still is that this is a "T Swiss Made T" dial (look closely to the left and right at six o'clock on the dial)—the luminous material is tritium based. Rolex phased out the use of the radioactive material entirely before the turn of the millennium.

A steel Daytona from this era is far more readily available than its rarer white gold cousin, especially given its Zenith-powered engine.

Good examples of white gold Rolex Daytonas were fetching a (more) modest $12,000 in 2010, but prices have almost tripled since then, not least because of the stunning auction success in 2017 of the Paul Newman "Paul Newman" Daytona ($17.75 million).

After five years of development, in 2000, Rolex launched a new Daytona with their own in-house movement (Calibre 4130), the first completely in-house movement in more than fifty years.

UNITS PER HOUR
ROLEX
OYSTER PERPETUAL
SUPERLATIVE CHRONOMETER
OFFICIALLY CERTIFIED
COSMOGRAPH
DAYTONA
T SWISS
MADE T

PATEK PHILIPPE
GENEVE
SWISS

The '90s neutral color palette ascended in a sort of anticonsumerist enlightened realignment of style that pushed back against '80s materialism and hyperconsumption. In interior design, the vibrant colors of the '70s and '80s morphed into more sullen colors. In contrast, and specifically toward the end of the decade, digital-based product execution became more colorful—consider 1998's Nintendo Game Boy Color and Apple's 1999 candy colored iMacs. While the '80s were still a tenuous time for the recovering watch industry, the '90s saw a firm revival in high-end watchmaking, especially in regard to watch complications. Following the fall of the Berlin Wall and the reunification of Germany, brands such as A. Lange & Söhne reconstituted and joined the watchmaking renaissance. Vintage watch sales spiked, and the dissonance of the quartz revolution faded into memory.

Omega gave its Seamaster a contemporary update and fitted Pierce Brosnan's James Bond with its quartz variation. This brand/character affiliation would last for many years. Rolex's Daytona with Zenith's El Primero movement was in production until 1999 (when it was replaced by Rolex's own in-house movement), and Panerai sales spiked following Sylvester Stallone's imprimatur and the significant screen time devoted to the brand's timepiece in the 1996 movie *Daylight*.[193]

Introduced in 1985, Patek Philippe's Calatrava (Ref. 3919J in 18K yellow gold) was a 1980s luxury dress-watch staple. One of the bestselling Patek Philippes of all time, it was a go-to watch for many Wall Street bankers.

Roman numerals, simple white glossy dial, hobnail bezel patterning—all Patek hallmarks, making for an icon of the '80s and '90s. The movement is the 215 PS, with the PS designation indicating *petite seconde*—small seconds.

Discontinued in 2006, it has a thirty-hour power reserve and a rather discrete 33 mm diameter.

193. Apocryphal stories abound regarding how Sylvester (Sly) Stallone came to be associated with the Panerai brand. Anecdotes suggest that he found the watch while shopping in Rome or Florence, but on the basis of photographic evidence of Stallone wearing a Panerai at least a year before his supposed purchase, it seems more likely that the brand's watches were gifted to him as part of a promotional celebrity endorsement.

The 2000s brought an outlandish amalgam of styles, sizes, and astronomical price points. With increasing use of personal computing technology and CAD/CAM manufacturing techniques, previous technical limitations disappeared. Extending into watchmaking were the extravagance and decadence exemplified by celebrity socialites such as Paris Hilton and TV shows such as *Sex and the City*.

Brands became bolder in their technological aspirations, and technology informed design choices. Watchcase dimensions grew enormous[194] and more complex, movements were given many added complications,[195] and previously unfamiliar materials became commonplace.[196] Seeing an uptick in demand for bespoke watches, master watchmakers[197] boldly collaborated with other watchmakers or ventured out on their own.

New brands proliferated. Time telling seemed to be secondary, if indeed necessary at all, and the timepieces of the 2000s and 2010s reflected this mindset with mechanical watches doubling as slot machines[198] or incorporating poker, baccarat, roulette, and blackjack.[199] The upper echelons of watchmaking considered timepieces much more as objet d'art than functional tool. The number of new timepieces offered to the public for more than $100,000 was unprecedented (even on an inflation-adjusted basis). Then the Great Recession hit.

2005/6 Omega Seamaster Planet Ocean 600M Co-Axial Chronograph 45.5 mm (Ref. 2218.50, MSRP: $7,200). A commanding presence on any wrist, and a watch that captured the quintessence of the midaughts design trend, the Omega Seamaster Planet Ocean released in several band variations (bracelet, water-resistant treated leather, or rubber strap), size variations, and dial and bezel color variations. With its color-coordinated pushers and sharp color schemes, it met with a warm consumer response. The model helped narrow the gap between Omega's and Rolex's dive offerings. The timepiece's in-house and long-awaited George Daniels innovation, the coaxial 3313, together with an impressive 600-meter depth rating and large in-vogue presence, made for a compelling choice.

Thomas Mudge's lever escapement has been around since ca. 1755, and despite hundreds of attempts, no escapement design had yet displaced it. Illustrious watchmaker George Daniels wrote the book on watchmaking, literally—it goes by the apt title *Watchmaking* and was originally published in 1981. It has since undergone numerous reprints and translation. Daniels's coaxial escapement (on which he had been working since the '70s) was quite revolutionary and difficult to industrialize. It took until 1999 before Omega commercialized a limited-edition version (calibre 2500) of the low-maintenance coaxial escapement. The first chronograph (base Frédéric Piguet) to feature the coaxial escapement was calibre 3313, which is installed in the pictured Planet Ocean 600M. In 2007, Omega released the fully in-house coaxial calibre, the 8500, which includes a silicon balance. Omega subsequently doubled their warranty to four years, acknowledging the movement's improved performance and reliability.

194. During the late aughts and early 2010s, the oversized trend was an example of a fashion that many, but not all brands adopted. Rolex increased case sizes on some of its models by a measly 1 to 2 mm—and then only several years after the trend took hold—while other brands embraced "bigness" almost immediately and created oversized models within many of their existing portfolios. The more recent vintage trend has similarly been adopted/rebuffed. U-Boat, an Italian brand founded in the mid-2000s, offered its U-1942 limited-edition watch in a gargantuan 65 mm diameter case.

195. Vacheron Constantin's Tour de l'Ile launched in 2005 to celebrate Vacheron's 250th anniversary and did so with sixteen complications in a limited seven-piece run of million-dollar-plus masterpieces. Patek Philippe launched the Sky Moon Tourbillon in 2001, with only a few manufactured each year at roughly $1.5 million each. In 2009, Jaeger-LeCoultre presented the Hybris Mechanica trilogy of complicated timepieces, including the 1,300-part movement and twenty-six complications of the Hybris Mechanica à Grande Sonnerie ($2.5 million). Also in 2009, Franck Muller's (19.15 mm thick) thirty-six-complication Aeternitas Mega 4 ($2.7 million) introduced a thousand-year calendar among its other twenty-five visible complications.

196. Please refer to the section on "Fashion" (within "Watch Styles") for additional insights.

197. Robert Greubel and Stephen Forsey, two highly acclaimed watchmakers, founded Greubel Forsey in 2004. F. P. Journe was founded by François-Paul Journe in 1999. He is the only three-time winner of the Aiguille d'Or grand prize from the Fondation du Grand Prix d'Horlogerie de Genève.

198. Girard-Perregaux's 2008 ($625,000) Vintage 1945 Jackpot Tourbillon offered a 125-combination, three-reel working slot machine operated by a handle on the case band.

199. Christophe Claret specializes in gaming timepieces, including these gaming complications.

OMEGA
Seamaster
CO-AXIAL
CHRONOMETER

Following the 2007–08 financial crisis, there seemed to be a tidal shift toward simplicity and the embrace of minimalism. The prolific launch of microbrands, many of them via crowdfunding, fast-tracked this development. The seeds of design restraint and advancement toward the 2010s vintage renaissance had support from Hollywood too. The period-perfect set design of 2007's television phenomenon *Mad Men*—beginning in the 1960s and depicting one of New York's most prestigious ad agencies and its antihero ad man Don Draper—was representative, if not determinative, of the cultural tipping point toward vintage that would take hold several years later. *Mad Men* spawned numerous copycat shows also set in the '60s, typically with uncomfortably tolerated misogynistic attitudes toward women, '60s gender-bending fashion influences, and the promise of a spacefaring future.

Richard Mille RM011-FM (first introduced in 2007). Automatic chronograph flyback (movement by Vaucher, flyback module by Dubois-Dépraz) with sixty-eight jewels and skeletonized dial. The FM designation is for "Felipe Massa" (the longest-serving sports ambassador for Richard Mille and distinguished F1 and Formula E driver). Massa assisted with the development of the RM009. The driver has been photographed wearing several Richard Mille creations. The RM011 is Richard Mille's bestselling watch and is still in production. The skeletonization craze of the aughts was given an assist by these hefty tonneau masterpieces.

RICHARD MILLE
TACHYMETER
START / STOP
RESET/FLY-BACK
RM011-FM
SWISS MADE

Coming out of the Great Recession (2007–09) meant that budget was primary in selecting interior design essentials and other durables. Unfinished metal and plywood and austere interiors resulted from a severe minimalism brought about by the protracted downturn and the growing international popularity of the KonMari method[200] (and derivative approaches), which encouraged people to hold on to only those things that gave them joy or meaning.

Tudor Heritage Black Bay (Ref. 79220R, ca. 2012, MSRP: $3,100). If one wristwatch exemplifies the enthusiastic vintage/heritage resurgence of the 2010s, it is the long-anticipated, instantly popular, and retro-styled Tudor Black Bay, first unveiled to the public at BaselWorld in 2012.

As opposed to a simple reissue, the Black Bay is evolutionary, incorporating elements from several previous dive watches under the Rolex/Tudor umbrella. Beating within is a Tudor-modified ETA 2824. The gilt dial, red bezel, domed sapphire crystal, and snowflake hour hand give it a distinctive vintage flair while still typifying the modern-minimalist trope.

While its predecessors exhibited a more modest 39 mm diameter, the Heritage Black Bay incorporates modern sensibilities for a larger upsized case (41 mm). Water resistance is 200 meters. The watch launched with a vegetable-tanned leather strap, steel bracelet, or woven fabric band.

200. The KonMari Method is the brainchild of Marie Kondo, the renowned Japanese organizing consultant whose method of organization sparked a Netflix series in 2019 and led to her book *The Life-Changing Magic of Tidying Up* (2011) being translated into several languages.

TUDOR
GENEVE
200m:660ft
ROTOR
SELF-WINDING
SWISS MADE
10
20
30
40
50

New watch brands and designs embodied this minimalistic attitude. The Apple Watch embraced "millennial pink" (ca. 2012, when its use exploded, although it originated earlier), a highly toned-down unisex version of Barbie pink[201] and the "it color" of the moment. The "millennial pink" trend has proven resilient. Hublot released a limited-edition, lightweight, anodized aluminum watch in 2020 with this blush shade incorporated into the strap and case.

At the dawn of the 2020s, vintage mania has firmly taken root. Offended by the vicissitudes of a luxury watch industry seeking outsized profits during the aughts (raising prices routinely[202]) and following the resurgence of a "mid-century modern"[203] design fascination, important watch aficionados and collectors moved their substantial wealth to acquiring vintage timepieces. The belief was that the pricing of vintage watches is determined by a free marketplace, not by "greedy" watch executives. Whether that is true of the auction market or not is a matter of debate, but the vintage watch market has certainly exploded.

The watch industry took additional note; after all, vintage reissues have always been a thing. But this time, the fascination with vintage spanned every decade of the twentieth century and provided manufacturers with additional opportunities to utilize their back catalog.

If the watch buyer's interests lay in vintage timepieces, then brands would eagerly satiate this demand. Watch manufacturers dusted off their old designs and gave them a twenty-first-century refresh. Thus, most recently, we have seen a surfeit of vintage-inspired timepieces from every decade since the 1920s, with the 1980s now coming increasingly under the spotlight. Breitling's newest Chronomat certainly hearkens to this decade.

The Chopard Alpine Eagle (Ref. 298600-3002, MSRP: $12,800). A major update to the St. Moritz of the 1980s, the Alpine Eagle provides Chopard with a viable alternative to the Patek Philippe Nautilus and Audemars Piguet's Royal Oak, both of which continue to command a premium in the luxury sports watch market. Shown here is the large model (41 mm) with galvanic-treated gray dial. There is also a smaller 36 mm diameter version and, as of late 2020, a chronograph version. An in-house chronometer-certified movement (01.01-C, 207 parts, thirty-one jewels) powers this watch.

201. Until the 1930s and '40s, pink was not yet well established as a "girl's color." In fact, as late as 1918, trade publications such as "Earnshaw's Infants' Department" supported pink as a decidedly boy-specific color. From the publication: "The generally accepted rule is pink for the boys, and blue for the girls. The reason is that pink, being a more decided and stronger color, is more suitable for the boy, while blue, which is more delicate and dainty, is prettier for the girl." Marketers prefer clear demographic delineations, and color is no exception. The more marketers can differentiate their product, the more they can sell. By the 1940s and certainly by the 1950s, manufacturers and retailers ensured that pink had transitioned to a girl's color. Its swing to being primarily a girl's color was likely fortified via the popularity of First Lady Mamie Eisenhower and her affinity for the color pink. In fact, the color of her ball gown for President Eisenhower's inaugural ball was officially called "First Lady Pink." The first lady was so popular that in 1952 she featured in Gallup's poll of "Top Ten Most Admired Women in America." Blue still reigns supreme, however, among both men and women. A 2015 YouGov survey puts blue as the most popular color for both genders. According to the same survey, blue is also the most popular color across the group of countries—ten countries were surveyed on four continents—usually followed by either green, red, or purple.

202. For reference, the Rolex Daytona 116520 (introduced in 2000 at a price of €5,179) was retailing for €10,950 in 2016. That represents almost 5 percent per year price inflation (compounded annually). Compare this with an average annual inflation rate of around 2.2–2.5 percent for the same period (for the US and EU, respectively).

203. Midcentury modern (MCM) is an American design movement that found popularity during the mid-twentieth century—an interpretation of the International and Bauhaus movements. The movement features clean lines and modern construction methods and was especially influential in residential design. It influenced architecture, graphic design, and industrial design from approximately 1945 to 1969, although it really became popular in the mid-1950s.

CHOPARD
CHRONOMETER
SWISS MADE

Bell & Ross
AUTOMATIC
SWISS MADE
27

As 2021 dawned and the lingering effects of COVID-19 persisted, a thorough revival and widespread adoption of late-twentieth-century "maximalism" was inevitable. Just like in the 1970s, when the neo-Expressionists pushed back against reductive minimalism, so too has a growing trend in the social fabric of the 2020s repealed minimalism and adopted its apparent antithesis. Design critics, however, are quick to point out that "maximalism" is so much more than minimalism's antipode. It is not consumerist per se, but about keeping and displaying objects that have meaning and place in our lives. In fact, these articles need not cost anything at all. Repurposed sidewalk "trash" can produce remarkable household ornaments. Think of it as eclecticism taken to the maximum. Our lives are so much richer with these items. They may clash with other things, but that is part of their delight.

The sparsely decorated, hard-wood-floored, 18-foot-ceilinged living rooms adorning *Architectural Digest*'s aspirational fantasies are just that: fantasies. Living in an increasingly complex world, forever changed by the COVID-19 pandemic, we have hunkered down in our homes and apartments, some of us doing so for many months, in habitats envisioned and curated for nocturnal sojourns when social distancing was not the norm. In 2020 and 2021, many people found themselves spending disproportionately longer stretches within their domiciles. Being able to display only one vase on a single coffee table next to the most suitable handpicked plush couch is no longer the calming influence it once was. These sparsely curated rooms may be beautiful to photograph and behold, but they are not truly livable spaces—at least not for most of us, and not now.

Before the pandemic, when we returned from our chaotic city jobs following a harrowing commute, the home, with its minimalist surrounds, cosseted us in silence and simplicity. Our craving for varied experience is no less formidable than it ever was, but it is exacerbated by long spans of time alone or with only a few others in a structurally limited space—usually the home. I predict, therefore, as part reaction to the socioeconomic tide brought about by the pandemic, a thorough global adoption of the eclectic and the handmade and a path toward maximalism. This design shift will manifest in watchmaking too. As pointed out elsewhere, watchmaking is not usually (or often) a leading indicator of design and traditionally takes years to catch up to burgeoning trends. Watchmakers have their work cut out for them, and I personally think that this is going to be the most exciting time in the history of watch design.

Bell & Ross BR05, launched at the tail end of 2019 (MSRP: $4,900), was immediately well received. The signature numerals and rounded case shape inherit stylistic cues from other Bell & Ross models, themselves inspired by aircraft cockpit instrumentation.

The integrated metal bracelet features alternating brushed and polished links. The twenty-five-jeweled BR-CAL.321, based on a Sellita SW-300, maintains a thirty-eight-hour power reserve. Water resistance is 100 meters. The watch shown here is the blue-dialed variation. Blue and gray versions were simultaneously released.

Clearly influenced by AP's Royal Oak and PP's Nautilus, the Bell & Ross design makes no apologies for what it is and what it aspires to be. The Vacheron 222, IWC Ingenieur, Girard Perregaux Laureato, Hublot Big Bang, and the Chopard Alpine Eagle all owe their familiar design vocabulary to the same AP and PP inspiration. In fact, the Hublot Big Bang is often compared to the Nautilus and Royal Oak. Ben Clymer, in a 2014 interview with Jean-Claude Biver, who was then head of LVMH's watch division, suggests that "many people say the Big Bang looks just like the Royal Oak offshore." Biver instantly retorts, "One hundred percent; looks like; so what? . . . because it's a similar idea . . . both took the porthole as an inspiration for the watchcase."

THE SMARTWATCH & DESIGN

The smartwatch can do almost anything "digitally" better than its mechanical analog predecessor. Watchmaking no longer requires a more accurate timepiece and can instead focus on materials science and state-of-the-art design. It will be interesting to observe how brands deal with the new "smartwatch crisis" in light of the gadget's ever-increasing global adoption.

Good design is no stranger to technology, and technology is much better off for it. We are all well acquainted with Apple's extraordinary obsession with design perfection across their product portfolio. Apple Watch's sleek cushion-shaped design looks deceptively simple—it's just another squircle (squared circle), right? Nope.

The Apple Watch CAD drawings below[204] and at right (used by third-party accessories designers to avoid covering sensors, among other things) illustrate the impossible radial complexity of what otherwise appears to be relatively simple curved shapes.

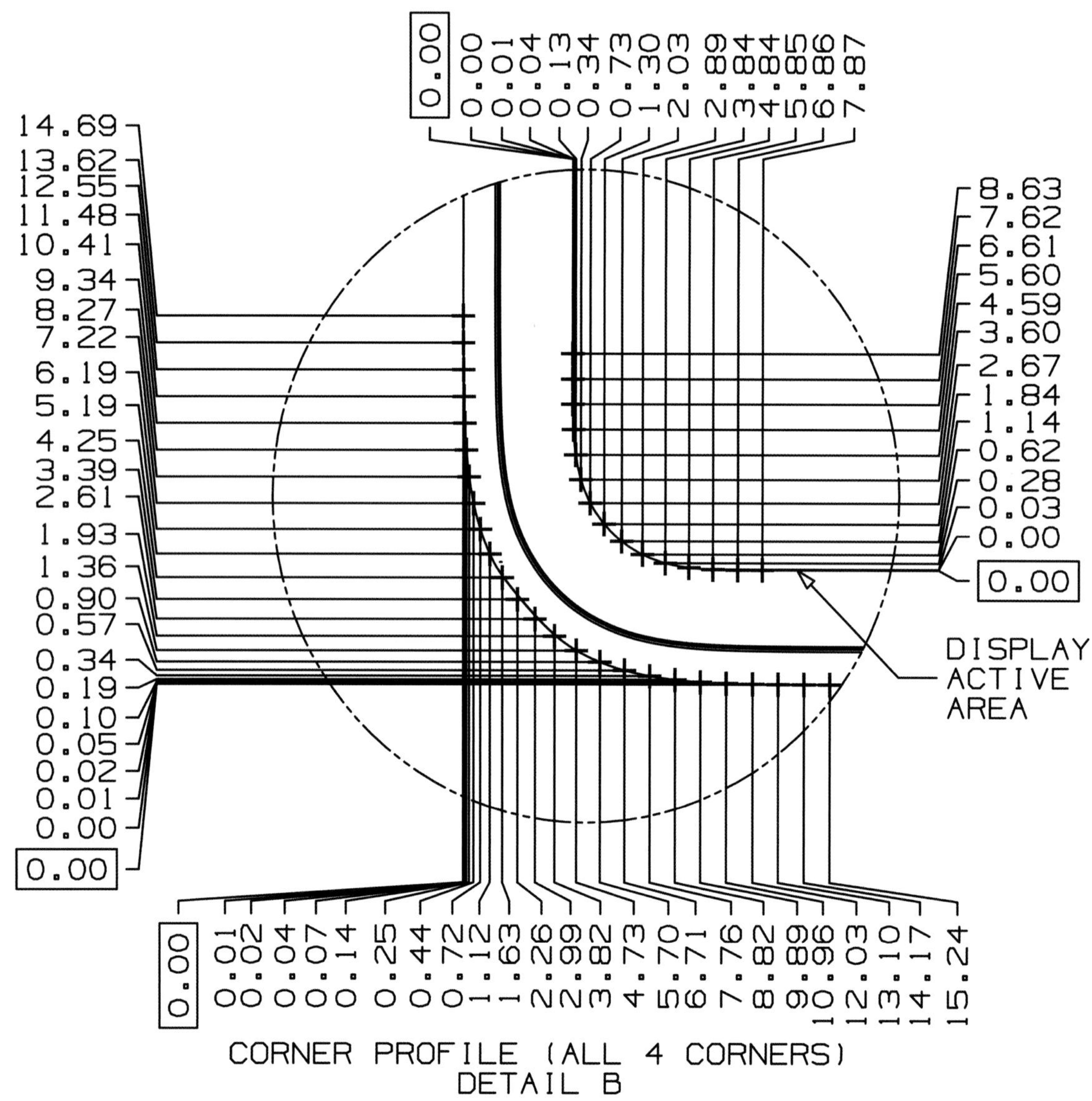

204. Source: Apple Inc.

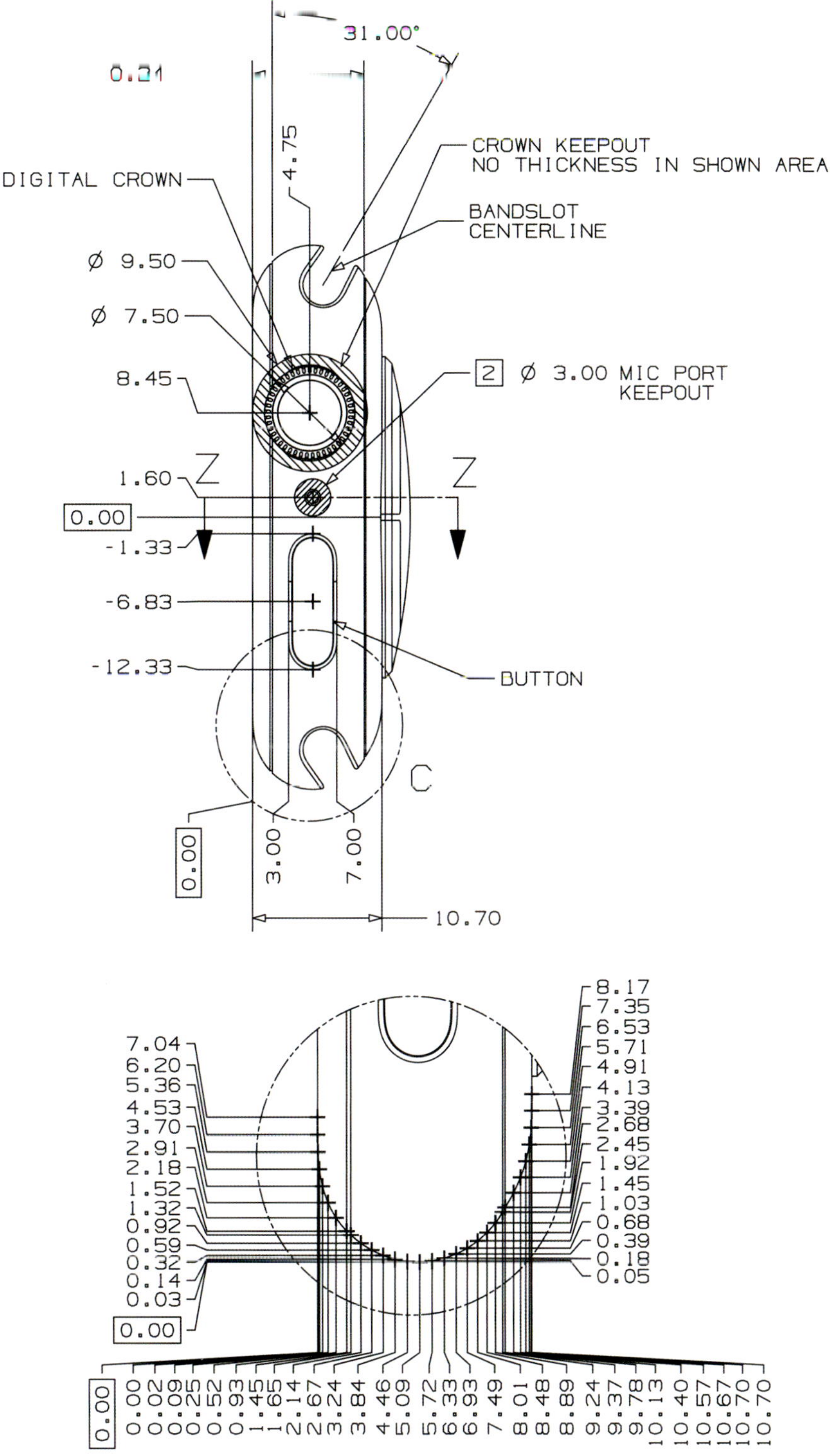

PROFILE (ALL SIDES)
DETAIL C

EVOLVING DESIGN: A CASE STUDY OVER FORTY YEARS

Within a particular brand's collections, it is often enlightening to examine a model's design evolution. It not only shows changing cultural tastes across years or decades but also focuses attention on a brand's adoption or avoidance of stylistic norms.

Several high profile wristwatch models that fit the criteria of evolved icons immediately come to mind: Rolex's Daytona and Submariner, Zenith's El Primero, Blancpain's Fifty Fathoms, Chopard's Mille Miglia, Cartier's Santos and Tank, Panerai's Luminor and Radiomir, Breitling's Navitimer and Chronomat, Omega's Seamaster and Speedmaster, TAG Heuer's Monaco and Carrera, and IWC's Portuguese and Jaeger's Reverso, among many other excellent iconic choices. All of these watches have a modern iteration that embodies the design motifs of its forerunners.

However, these models have evolved their designs (for the most part) cautiously and iteratively to the extent that it is often difficult to see changes from one model generation to the next or even across several generations or model years.

A watch design that perhaps shows a more radical historical progression across generations—and is arguably equally justified in holding a place in modern horological history (along with the Nautilus and Royal Oak) —is the Vacheron Constantin Overseas. It began its design journey as model 222 (although one could possibly go back even earlier, to 1975's 36 mm reference 2215 Royal Chronometer, for a progenitor featuring an integrated bracelet, albeit with an octagonal case).

THE 222

Not to be overshadowed by Patek Philippe and Audemars Piguet with their revolutionary Nautilus and Royal Oak sports models, Vacheron Constantin introduced its own sports watch: the 222 in 1977. Jörg Hysek, abstracting several design cues from the aforementioned Gerald Genta, pioneered this tonneau-shaped sports watch for Vacheron in celebration of its 222nd year. Incorporating a bold gold inlay of Vacheron's iconic Maltese cross logo, [205] only five hundred units of the now highly collectible[206] model 222 (reference 44018) reputedly left the factory in steel. One hundred twenty 222s came in two-tone steel and gold, and a further one hundred in an all-gold version.

Admittedly, these timepieces did not initially fly off the shelves, and thus many 222s feature registration paperwork dated several years following their production dates. The watch appeared in 34 mm and 37 mm varieties.

Vacheron Constantin 222 (Ref. 44018/411) with 37 mm case diameter (ca. 1980). It shares an ébauche[207] with Patek Philippe's Nautilus and Audemars Piguet's Royal Oak (sourced from Jaeger-LeCoultre).

205. A heraldic cross developed from earlier forms of the eight-pointed crosses from the sixteenth century, consisting of four converging V or arrowhead shapes.
206. Clean examples of the 222 currently trade for around $30,000.
207. An incomplete watch movement lacking a regulating organ (balance, hairspring, escape wheel, lever, and pallet stones or jewels), dial, and hands.

VACHERON CONSTANTIN
GENÈVE
12
AUTOMATIC
SWISS

THE OVERSEAS

The successor to the 222, the Overseas launched at the tail end of 1996. Just as elegant, dressy, and thin as the 222, it consisted of 35 mm and 37 mm case sizes, with a smaller women's model.

Developed by outside designers including Dino Modolo, and in conjunction with Vacheron's Vincent Kaufmann, it found immediate success. Additional dials and a chronograph version were introduced over time until the second-generation Overseas appeared in 2004.

Vacheron Constantin Overseas I (ca. 2001, Ref. 42042/423A, in 37 mm case). The bezel superbly incorporates Vacheron Constantin's Maltese cross into the overall design.

ACHERON CONSTANTI
GENÈVE
22
CHRONOMETER
AUTOMATIC
SWISS
MADE

THE OVERSEAS II

The Overseas II evolved the design more substantially than the previous model alteration. The case size increased to 42 mm, and the bracelet began integrating the Maltese cross into its links. The bezel also implemented the Maltese cross motif carried over from the Overseas I. Unlike the Overseas I, which offered a chronograph complication only as an alternative, the Overseas II model family offered a dual time complication and a limited series of perpetual calendar timepieces in addition to the standard timepiece.

Vacheron Constantin Overseas II (Ref. 47040/000A, ca. 2015, MSRP: $13,000) produced in a limited series of 350 pieces with a striking blue dial. Both the bracelet and bezel incorporate the Maltese cross motif.

VACHERON CONSTANTIN
GENEVE
AUTOMATIC
ANTIMAGNETIC
SWISS
MADE

THE OVERSEAS III

In 2016, Vacheron Constantin unveiled its third iteration of the Overseas—the fourth iteration of an evolving design if you include the 222, or the fifth iteration if you include the 2215 (a stretch, I think).

The third series of the Overseas further refined the design and introduced a time-only model (Overseas Self-Winding), an ultrathin version of the same (in white gold sans date), and an ultrathin perpetual calendar. The current model incorporates a tool-free quick-swap strap/bracelet system that makes swapping out bracelets and straps a doddle, but it unfortunately limits the use of after-market bracelets and straps.

The Overseas II's Maltese cross motif in the bracelet and bezel carries over but appears a little less pronounced in the new bracelet layout. The gorgeous blue dial encompasses several subtle shades of the hue. The power reserve of the contemporaneously introduced calibre[208] 5100 (for the non-ultra-thin version) extends to sixty hours, and the movement features Geneva seal hallmarking.

Vacheron Constantin Overseas III (Ref. 4500V/110A-B128, MSRP: $19,600, 41 mm) in stainless steel. The case back is transparent, providing the wearer with a gorgeous view of calibre 5100's 172 parts. The watch is emblematic of the late 2010s general stylistic refinements in terms of smaller case size, less ornamentation, and simplified dials.

208. "Caliber" and "calibre" are different spellings of the same word. Caliber is the preferred spelling in the US, and calibre is preferred in all other native English-speaking countries. The word derives from the French *calibre* and came to English in the sixteenth century. Interestingly, "caliber" with the -er ending was used more than twice as often by British English speakers in the mid-1700s than its French-derived antecedent. By 2000, British English speakers used calibre almost four times more often than the alternative. In the US in 2000, caliber surfaced almost seven times more often. Regardless of spelling, caliber or calibre is synonymous with movement. As with gunmaking, it designates diameter. I have chosen to use "calibre" in all instances.

VACHERON CONSTANTIN
GENEVE
31
SWISS MADE

VACHERON CONSTANTIN
GENÈVE
AUTOMATIC
SWISS
VACHERON CONSTANTIN
GENÈVE
CHRONOMETER
AUTOMATIC
VACHERON CONSTANTIN
GENEVE
AUTOMATIC
ANTIMAGNETIC
SWISS
MADE
VACHERON CONSTANTIN
GENEVE

Evolving design: four generations of Vacheron Constantin's sports watches. Beginning with the 37 mm 222 from 1977 in the upper left corner, the first generation Overseas from 1996 is at top right, the second generation at bottom left, and 2016's third generation (with more subtle styling cues and 41 mm case size) at bottom right.

DECORATIVE WATCHMAKING TECHNIQUES

Decorative watchmaking techniques are too numerous to mention in a book of this length or focus, but I have tried to include a few of the more common and interesting ones. Ironically, the initial purpose of most of these decorative techniques was not (at least primarily) decorative—it was utilitarian. In addition to the artisanal beauty of the assorted crafts involved, many of the techniques listed below slowed or reduced entirely the potential for corrosion and oxidation of the metal parts or kept surfaces free of dust and other particles.

Bluing of screws, superficially ornamental, is actually a consequence of raising the temperature of the metal to harden it. As the part heats up, its color changes, until it turns to the desired blue shade. While some modern movement makers use painting or electroplated bluing, the archaic yet practical screw-heating process achieves dual goals—it is decorative and functional.

Modern decoration and electroplating are much the same as they were decades or even centuries ago. Today, however, CNC machining supplements or replaces many of the processes. The decoration of the movement and its parts adds a premium to the value of the movement and to the overall watch. As a rule, the greater the degree of artisanship and effort, the higher the final price of the finished watch. Many of the techniques have a machined equivalent, but the hand-finished part still reigns supreme.

Brushing (brossage): A fine or very fine brushed or satin finish applied via a rotating wheel

Chamfering (*anglage*) or "polishing angles": polishing a chamfer of a previously contoured section. A chamfer is a transitional edge between an object's two faces—sometimes called a bevel. In machining terminology, chamfering refers exclusively to beveling for safety purposes by removing sharp burrs left over from previous machining processes.

Aside from looking beautiful, chamfering also reduces the potential for oxidation. The edges of the plates shown on the back of this movement display mirror-like (*poli miroir*[209]) chamfering, generating incredible depth as the light bounces off the surfaces.

Anglage requires application of several metal files to the work piece with diminishing abrasive strength, followed by superfine polishing with peg wood and diamond paste.

Corners and sharp angles are particularly difficult and time consuming to hand-polish. Exceptional anglage is a hallmark of only the finest watchmakers, who will ensure that the polished edges are uniform throughout the part.

Similar to *anglage*, but using CNC, diamond planing (*rabotage diamant*) creates a highly polished chamfer by using a nonrotating diamond bit. Performed correctly, it achieves the same effect as *poli miroir anglage*.

209. Also called poli noir. When the perfectly polished, blemish-free object is perpendicular to the viewer, the surface appears black.

Engraving (*gravage*): Removal of material to create engraved images (sometimes personalized for bespoke timepieces), logos, manufacturer's marks, calibre, hallmarks, or other lettering; with or without varnish infill,

CNC machinery and lasers perform the majority of modern engraving tasks, especially where large batches of timepieces are concerned.

Carving/chiseling (*ciselage*): Removal (typically using beveled chisels) or repelling of material (with a hammer). Usually performed only on complicated or specialized pieces and then only by master artisans. Hydraulic presses can achieve (somewhat) similar results to the artisanal method, but the finished part lacks the depth, uniqueness, and finishing of a hand-chiseled design.

The image above also illustrates *l'ajourage* (openwork or skeletonization) on the rotor, revealing the beauty of the movement below. The artisan makes small holes in the work piece and then inserts a tiny saw blade or cutter to make the openwork cut out.

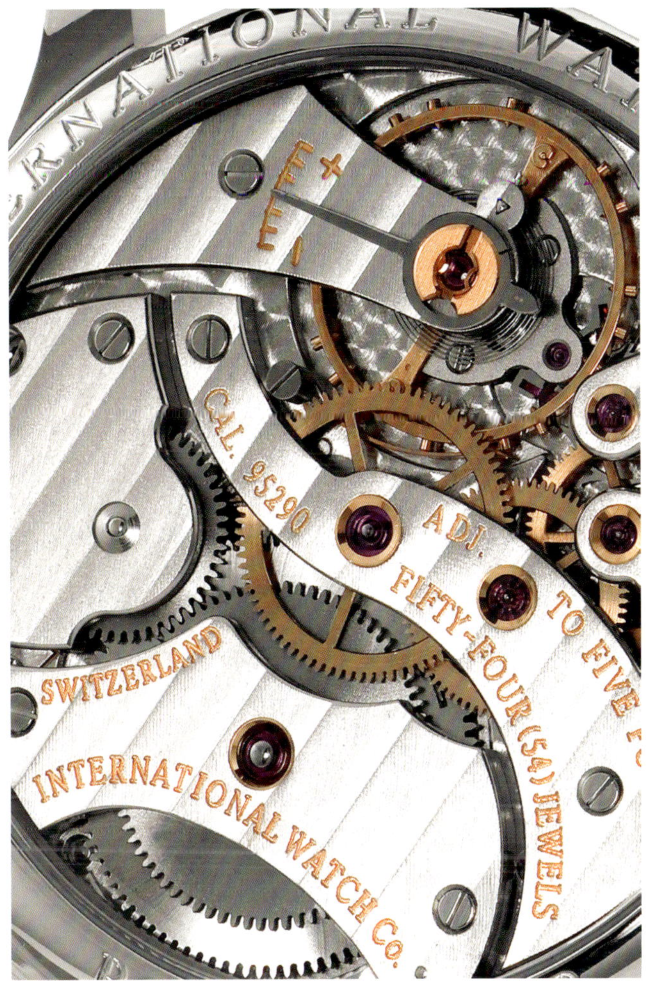

Geneva stripes / waves / straight ribs (côtes de Genève, vagues de Genève, or *côtes droites*): The creation of parallel lines or curves at regular intervals, using an abrasive tool (usually a wooden burnisher with applied abrasive material).

Because of the relatively substantial amount of material removed, and given the extremely tight tolerances in watchmaking, it is a purely decorative technique used mainly on movement bridges. When performed with CNC or other machinery, it is termed "mechanical ribbing" or "côtes mécaniques." CNC machining performs this task well and uniformly, and usually in a single pass, but lacks the uniqueness of handmade stripes.

Circular graining (*côtes circulaires*): Regularly spaced circular polishing using an abrasive tool. Similar to standard côtes de Genève but with a circular arc.

Smoothing (*adoucissage*): Variable linear surface decoration made using an abrasive paper or emery cloth

Spiraling (*colimaçonnage*): Spiral surface decoration obtained by rotating the abrasive grinding wheel on the part surface

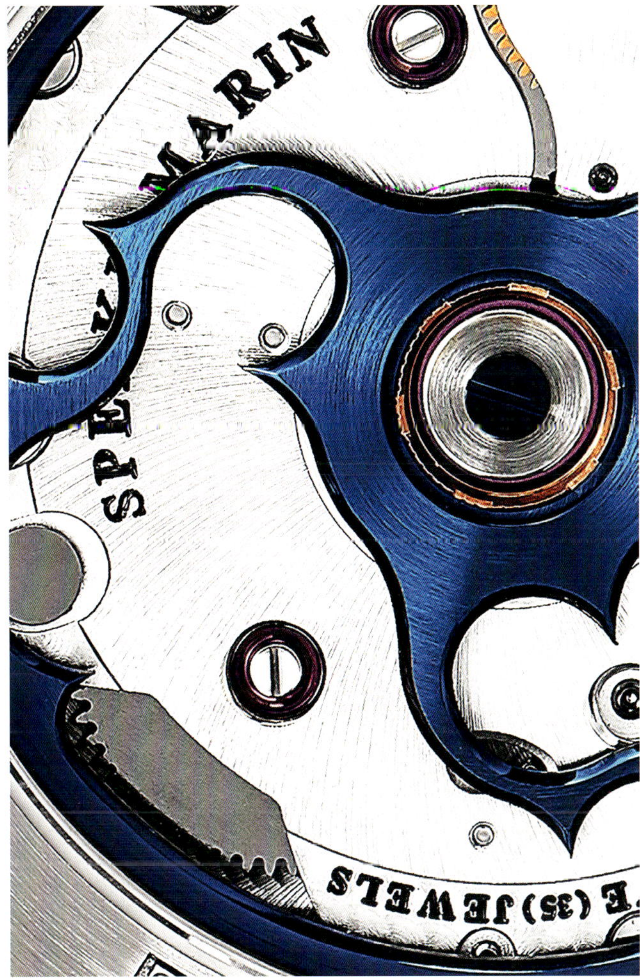

Sunburst (*le soleillage*): Outward radiating lines or curves from a central intersecting point. The part and polishing wheel move in opposite directions.

Beading (*perlage* or *le perlage*): Successive (often overlapping) circular patterns made using a rotating abrasive rod; mainly utilized for plates and bridges but occasionally adorning dials. The size of the swirl may differ depending on the part. In addition to preventing corrosion, the repeating circles create a beautiful swirling pattern.

Sandblasting/microblasting (*sablage*/*microbillage*): Bombarding the surface material (the watchcase in this example) with abrasive microspheres (of metal, glass, or ceramic)

Shot peening (*grenaillage*): Similar to sandblasting, the process itself strengthens the part and relieves stress by spreading the material "plastically" rather than through a method of abrasion. It prevents the propagation of microcracks from the surface, which in turn helps reduce oxidation and corrosion.

Each wire bristle of the specialized brush functions like a micro ball-peen hammer (hence peening). Less material is lost in the *grenaillage* process than in the *sablage* or *microbillage* process, and the surface appears more radiant. A related process, known as peen plating, can apply material directly onto the surface.

The *grenaillage* treatment process traditionally involved the use of a mercury/gold slurry to coat an aluminum-powdered brass plate and then evaporating the mercury. Due to mercury's toxicity, the process no longer utilizes the harmful metal.

First, as with the Urban Jürgensen dial shown here, the artisan gently engraves the surface. Black lacquer fills the engraved recesses for the indexes and numbers. Then, with very fine diamond paper, the markers are sanded down; the dial is brushed with metal bristles and coated with a silver-powder salt compound, creating an electrochemical reaction and producing the desired effect.

Guillochage (guilloche): Mechanical engine-turned patterns. Using a manually operated turning machine (a.k.a. rose engine), an architectural design of curved interlacing patterns is engraved into the material—an exceptionally time-consuming and intricate process. The Urban Jürgensen dial shown here exhibits two distinct patterns using this very old technique: one for the main dial (*grains d'orge*—barley grain pattern), the other for the small second subdial (*décor panier*—basket decoration).

According to the manufacturer, these patterns require as many as seven hundred operations and two days of handwork to complete. CNC *guillochage* is similar to rose-engine guilloche but instead utilizes a CNC machine; using a variety of decorative bits, decorations are incised on the surface of the material.

CNC *guillochage* performed by a competent operator is almost indistinguishable from rose-engine examples, but it is difficult to personalize for small batches. Less expensive watches utilize dials with hydraulically stamped patterns that mimic guilloche in terms of pattern but lack the radiant depth that only a rose-engine or mechanical tool can produce.

Rose engines are hard to come by, with the number of people who can operate them, or repair/restore them, fewer still. Maintaining proper chisel depth and flatness of the machined surface, which can be almost any engravable material, requires extreme patience and exquisite skill, followed up with careful polishing with finishing paste to remove tool marks.

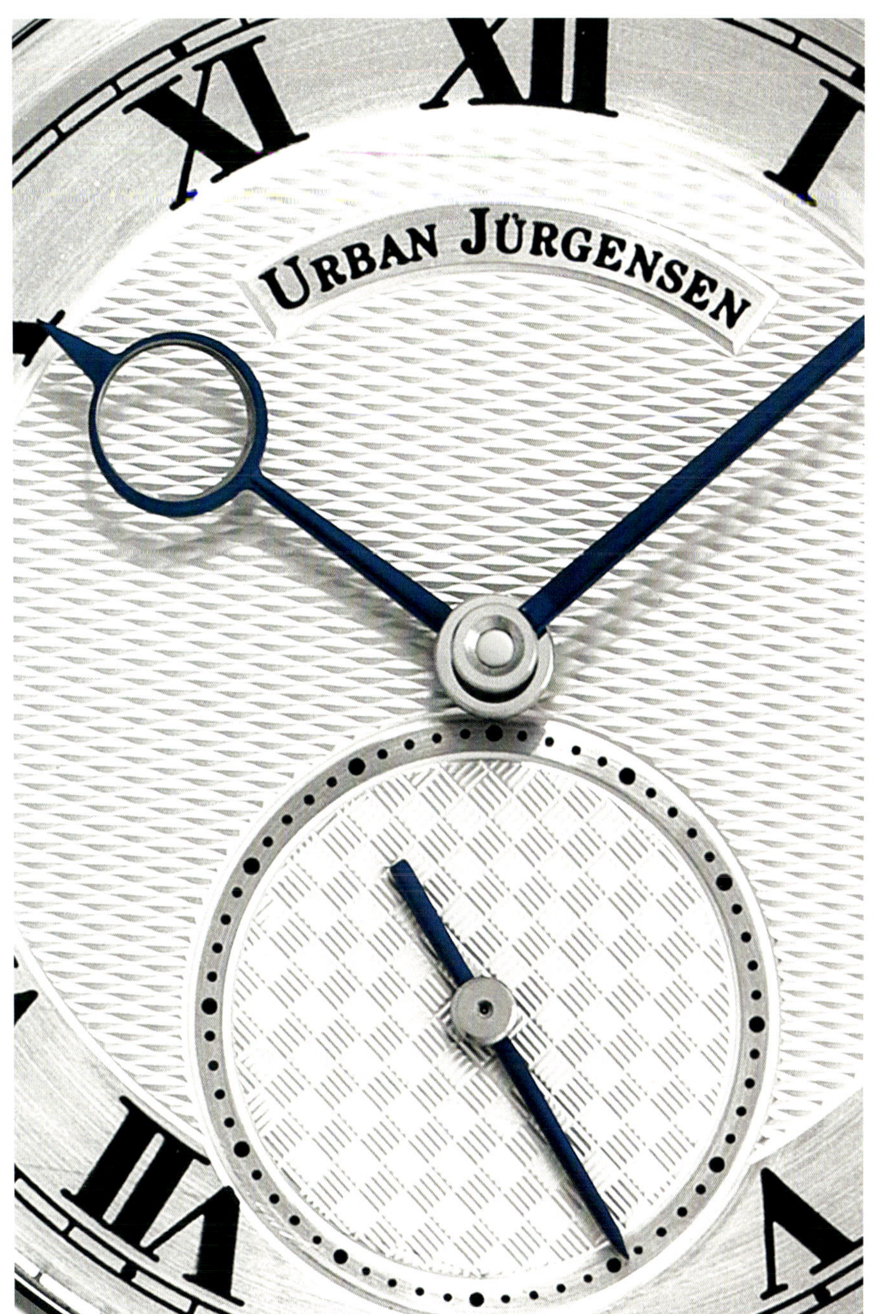

"Time is not just the hands of a clock.
It isn't just a change of seasons either.
It is a treasure trove of precious moments,
infinite little joys and blessings, precious
smiles, tears and heartbeats."

–Mona Soorma, *Soul Food and Instant Karma*

DIVE/NAUTICAL

The dive watch is a watch designed for underwater diving. Modern technology has allowed for the creation of dive watches designed to go as deep as 15,000 meters (approximately 49,213 feet).[210] A true modern diver's watch harmonizes with the ISO 6425 standard, which defines test standards and features for watches suitable for diving in depths of 100 meters (330 feet) or more. The personal dive computer has largely supplanted the diver's watch, but for many, the watch serves as a reliable backup device. Perhaps the most famous extant diver's watch is the Rolex Submariner, first introduced to the world at the Basel Watch Fair in 1954.

Watchcases for diving watches must adequately resist water pressure and endure the galvanic corrosiveness of seawater. To ensure this, the cases generally comprise materials such as 316L or 904L stainless steel and other strong alloys. To withstand a seawater environment at depth, watchcase construction for dive watches is sturdier than in a typical dress watch. Consequently, diving watches are relatively heavy and large compared to dress watches made out of similar materials.

Using the unidirectional dive bezel: The time shown above is 10:10[211]. The Rolex Submariner bezel was set to zero just prior to the dive (*purple arrow*) by rotating the bezel counter-clockwise. The "zero marker" is the blue triangle with white "pip" (shown at eight o'clock) on the bezel. At the dive's commencement, the time was 9:40 (thirty minutes earlier). Assuming the diver has only thirty minutes of breathable air, he or she needs to know when that air will expire; of course, today we have dive computers that perform this task, but the dive watch acts as a consistent and reliable backup. When the minute hand (which was at the eight o'clock blue-triangle position at submersion) reaches the two o'clock position (red minute hand, thirty-minute red marker on the bezel), a full thirty minutes has elapsed and the air supply is presumably exhausted. Time to surface.

210. In 2019, the Omega Seamaster Planet Ocean Ultra Deep Professional became the deepest diving watch yet created.

211. A common setup for watch photography, because it makes the hands look symmetrical and in the form of a "smile."

Jaeger-LeCoultre 185T470 Master Compressor Diving Pro Geographic ($27,500 when new in 2009). 46 mm titanium case with mechanical-analog 80-meter depth gauge (*blue arrow*) and world time.

The depth gauge features a novel membrane attached to the left flank of the case. Unlike other systems that can clog due to water exposure, this particular innovation connects the membrane to a specially calibrated spring and wheel system and does not require water exposure for operation.

Jaeger-LeCoultre leaned on decades of experience garnered from the production of their Atmos clock in designing this movement.

Analog dive watches most often feature a rotating bezel, providing easier read-off of elapsed time of less than one hour. The bezel computes the length of a dive. Just before or upon entering the water, the diver aligns the zero on the bezel with the minute hand. He or she then reads the elapsed time from the bezel instead of remembering the water entry moment and then performing a calculation of elapsed dive time. On dive watches, the bezel is unidirectional—that is, it contains a ratcheting device usually calibrated to 60 or 120 clicks. Being unidirectional, the bezel turns counterclockwise only to *increase* the apparent elapsed time, should the bezel be unintentionally rotated during the dive. If the bezel could turn clockwise, it may suggest to a diver that the elapsed time is shorter than reality, thus indicating a falsely short elapsed time reading and therefore falsely short saturation period—an assumption that can be extremely dangerous, on the basis of the limited supply of breathable air. Some dive watches feature a locking bezel to minimize the chance of unintentional bezel operation underwater. The bezel typically features a luminous "pip" or marker at the zero point and may, in addition, also feature five-minute or even one-minute increment markings.

Dive watches also typically feature a specialized crown protection mechanism, either with additional gaskets, a screw-down feature, or a protective locking device.

PAM00311 features Panerai's patented locking bridge device, which helps secure the watch crown and ensure optimal water resistance.

OFFICINE PANERAI
SWISS MADE
EIGHT DAYS
THREE

Some dive watches intended for saturation diving at great depths are fitted with a helium escape valve to prevent the crystal from blowing out due to internal pressure buildup. This buildup is caused by helium that has seeped into the watchcase in helium-enriched environments—*helium atoms are tiny, the smallest natural-gas particles found in nature*—as the watch and diver adjust to normal atmospheric conditions following a saturation diving event.

Many dive watches are fitted with some form of expandable bracelet or strap to accommodate the added circumferential dimensions of a wetsuit sleeve. In addition, as the diver descends, the compression effects of the increased water pressure serve to shrink the overall circumference of the wrist. Some wristwatches such as the Tudor Pelagos feature an automatic microadjustment mechanism that expands and contracts the clasp to ameliorate these compression effects. The NATO-style strap, which has recently proved popular on fashion watches, was designed (and is part of a military specification) to prevent loss of the wristwatch should a spring bar fail during usage. This one-piece fabric strap slides under the watch through both spring bars.

Underwater, low light conditions compromise legibility. To improve clarity, watch manufacturers typically enlarge the hands and hour markers and coat them in some form of luminous material such as luminous pigment or self-powered gaseous tritium light sources encased in thin glass tubes.

Ball uses applied tritium tubes for hour and hand markers. If you look closely, between seven o'clock and eight o'clock, the dial is marked T25, indicating the use of tritium (not exceeding 25 millicurie). Coincidentally, tritium's half-life is almost half of 25, at 12.32 years. Thus, the intensity of these tubes will reduce by half every approximately 12.5 years. Eventually, the tritium will become too dull to see and will necessitate replacement to restore its intensity.

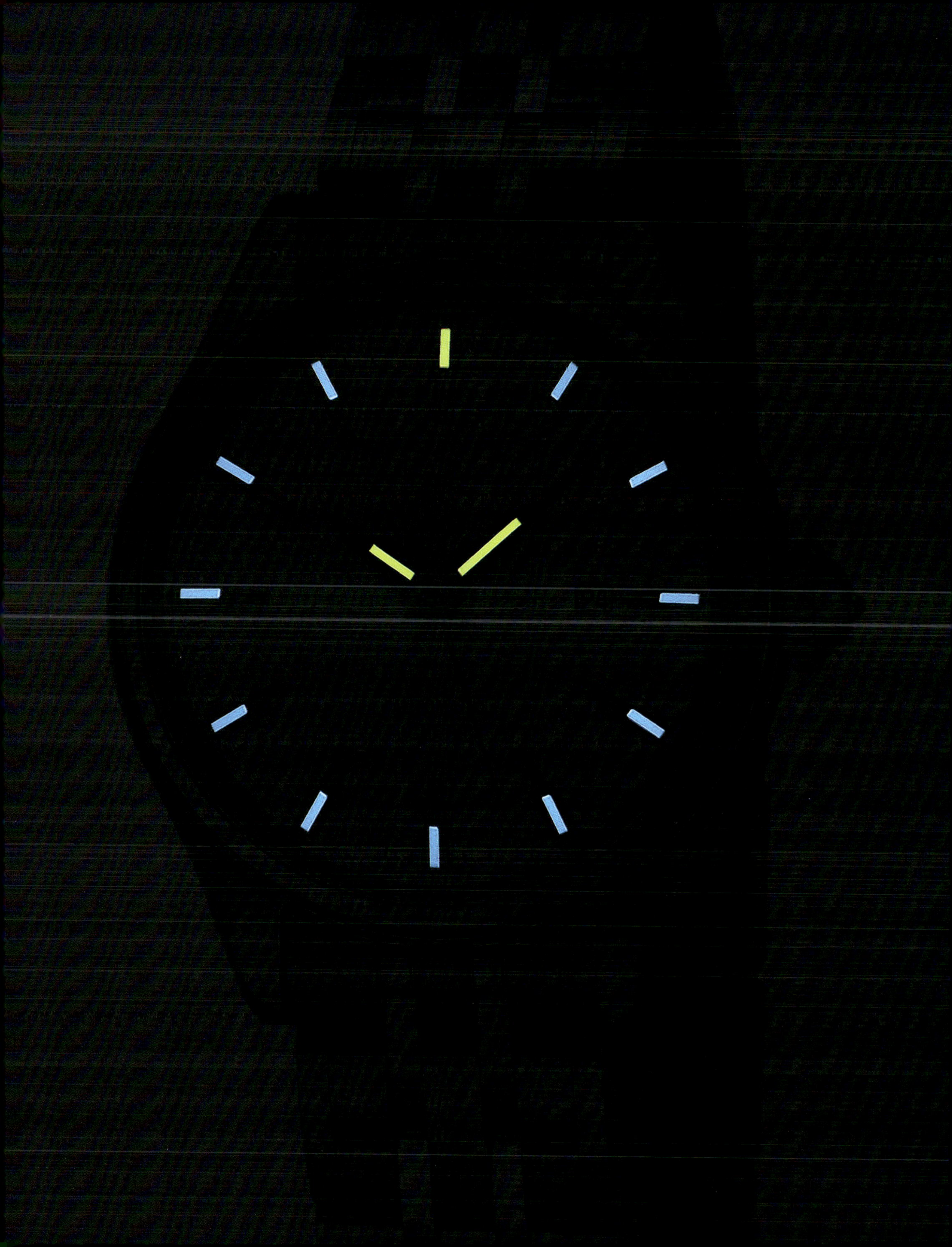

Historically, watch manufacturers applied radioluminescent paint (invented by Sabin Arnold von Sochocky in 1908) to the dials and hands of their watches. Radioluminescent paint will glow until the radioactive isotope has decayed.

The original radioluminescent paint incorporated radium-226 (together with zinc sulfide). Radium-226 is a radiological hazard (with a 1,600-year half-life) emitting gamma rays that can penetrate the watch dial and seep into human tissue. In fact, the high-profile legal case of the "Radium Girls"[212] resulted in the banning of radium for this purpose, and it was gradually phased out of use.

Though promethium was a suitable alternative to radium, its relatively short half-life of only 2.62 years made it an impractical solution for watch dials; it would lose half its brightness in less than three years. Tritium is the most common extant radioluminescent material, with a half-life of 12.32 years (see above), but since it is also radioactive (although far less harmful than radium), it requires special handling and application, changes coloration with time, and today is found only in new watch dials in the form of applied encapsulated glass tubes.

If one were to examine older watch dials, those with tritium-based paint will be indicated as such with the markings "T Swiss T" or "Swiss T < 25" highlighted on the dial, usually at the six o'clock position. Because it is highly regulated, modern tritium tubes are far less common than photoluminescent paint (such as Super-LumiNova), which glows in the dark after light exposure. While suffering from the disadvantage of not being self-powered (like radium or tritium), these afterglow pigments are a much safer alternative to radioluminescent paints, don't fade with time, and are now a staple in the watch industry.

A few brands, most notably Rolex, have introduced their own in-house versions of Super-LumiNova (now available in a rainbow of colors). Rolex calls theirs Chromalight, which glows with a distinctive bluish hue.

The Yacht/Regatta timer also falls into the nautical category. The most popular yacht timer is probably the Rolex Yacht-Master. It provides a ten-minute synchronized countdown used prior to the yachting event as the boat maneuvers to the start line.

FASHION

As the term implies, fashion watches follow current trends and styles of dress. The millennial-driven trend to simplify, reduce clutter, and focus on "experiences rather than things" heavily influenced the 2010s. Many nascent watch brands have exploited this trend with minimalist modern timepieces sometimes featuring nothing more than a thin steel case, hour and minute hands (no second hand), and a basic leather or NATO strap.

The late aughts introduced a trend that garnered the moniker "boyfriend watches." These are watches that a man would typically wear, but worn instead by a woman. The trend appears to be waning but is still somewhat popular.

The "bigness" trend that materialized in the midaughts (but is now in steady decline) had average watchcase sizes increase from 38 to 44 mm (men's watches) and beyond, with some timepieces bordering on 60 mm diameter cases.

Franck Muller's Vanguard Carbon with carbon fiber case and hand-finished art deco raised numerals. Ref. V45 SC DT CB NR (ca. 2015). The distinctive tonneau case shape fluctuates in popularity, but it offers a unique look to the watch connoisseur wanting a wristwatch that is neither circular nor rectangular. Franck Muller wristwatches are easily identified by their "Cintrée Curvex" case shape, first introduced in 1992 and curving in three dimensions—something not seen previously.

212. The "Radium Girls" were a group of women in the 1920s and '30s who painted the dials of watches using paintbrushes dipped in radioactive paint and who subsequently suffered severely adverse health effects.

330°
NORTH
300°
WEST
240°
GENEVE
VANGUARD
SWISS
MADE
210°
SOUTH
150°

Since about 2010, vintage watches have experienced substantial interest as throngs of collectors, connoisseurs, and opportunists have entered this market in search of a limited supply of timepieces. Some wear these watches because of nostalgia; others to differentiate themselves from the "run-of-the-mill" Rolex, TAG Heuer, Breitling, or Patek Philippe.

Blue-dialed watches seemed to be a fad but have held remarkably steady in popularity[213] and are now almost on par in sales with black- or white-dialed timepieces. Toward the end of the 2010s, green dials made a strong showing and continue to grow in adoption in the early 2020s.

Whereas yellow gold was the most popular precious-metal option in the 1970s, today the avid watch enthusiast can often choose from the same model in a variety of precious metals, including rose gold, white gold, platinum, and palladium. Several brands, including Rolex, Patek Philippe, Omega, and Hublot, have introduced their own form of colorfast rose gold to counter the effects of the standard alloy, which is subject to discoloration when the copper content within the rose gold oxidizes.

Dive watches, once the exclusive domain of the serious diver, adorn the wrists of many who have no intention of diving, let alone swimming, with their timepieces. Dive watches are de rigueur in some circles on Wall Street, and the slightly derogatory term "desk diver" is to indicate watches that have never been submerged in a large body of water.

The NATO strap, also known as the G10, is a strap originally requisitioned by soldiers in the British military through the completion of a G1098 form. It hearkens back to 1973 and is a nylon strap designed to incorporate a fail-safe should one of the spring bars between the lugs break or dislodge. Nowadays, watches across the spectrum, from entry level to Haute Horlogerie, feature this ubiquitous strap.

Prior to the advent of modern alloys and polymers, stainless steel and precious metal were the only typical materials found in watchcases or bracelets. Today, carbon fiber, aluminum, titanium, ceramic, plastic, wood, sapphire crystal, and a host of other modern materials, polymers, and composites make up these parts.

Precious gemstones, diamonds, crystals, and other jewellike adornments, once used only sparingly (typically on the bezel of a watch or as markers on the dial), now decorate entire watchcases and bracelets, blinged out with diamonds, precious stones, or Swarovski crystals.

Color palettes for watches are extensive, and every year watchmakers try to outdo each other with the introduction of new color combinations for the case, bracelet, dial, strap, or decorative embellishments.

For standard stainless-steel timepieces, newer methods of coating watch parts, including physical vapor deposition (PVD) and diamond-like carbon (DLC), have allowed manufacturers not only to beautify their watches but to toughen the finish too, using sophisticated high-pressure and high-temperature electroplating processes.

Dial materials, once constructed from only easily workable brass, now feature scores of other materials, including mother-of-pearl, enamel, feathers, insect wings, marquetry, meteorite, nylon, and denim, which are usually but not always affixed to a brass baseplate.

Bvlgari Serpenti Spiga spiral watch Ref. 102532 (ca. 2016) with 35 mm 18K rose gold and black ceramic case and black lacquered dial. Bezel inset with thirty-eight brilliant-cut diamonds. Crown set with black ceramic cabochon.

213. Numerous international studies routinely show blue as the favored color of both men and women.

SWISS MADE

French designer Christian Dior, in collaboration with industrialist Marcel Boussac, formed the now-iconic fashion house in late 1946. Success came quickly for the capable forty-one-year-old fashion designer. From designing for leading Hollywood stars to the cover of 1957's *Time* magazine, Dior changed the face of fashion.

The Dior Grand Bal collection (dating to 2011) takes inspiration from the beautiful haute couture ball gowns designed by Dior himself. The front-facing oscillating weight, to which the feathers adhere, is not just decorative but serves as the rotor for the winding mechanism. The feathers, sapphires, and lacquered dial with complementary coloring present a harmonious and radiant triadic color scheme.

Dior has combined golden threads, mesh, silk, feathers, precious, and semiprecious stones and even beetle wings into its kaleidoscopic designs for the Grand Bal collection.

The Grand Bal shown here in polished steel with diamonds is 36 mm in diameter (Ref. CD153B12A001, MSRP: $28,800). The opaline ring adjacent to the circle of diamonds is white mother-of-pearl. A transparent magenta-tinted sapphire crystal covers the case back.

If an off-the-shelf design such as this one is not to your liking, Dior also provides its clientele with a bespoke service to customize the Grand Bal. There are two hundred million possible combinations, with prices ranging from about $30,000 to $190,000. Custom timepieces take five months to produce and assemble.

Dior

Breguet Classique Automatic
(Ref. 7147BB/29/9WU, MSRP: $21,500, ca. 2017)
40 mm men's dress watch (6.1 mm thick) with magnificent white grand feu (big/great fire) enamel[214] (emaille) dial featuring painted minute track and numerals. The case is in high-polish 18K white gold with standard Breguet vertical case band fluting.

The movement, viewable through the transparent exhibition case back, is the thirty-five-jeweled Calibre 502.3 SD (12 lignes and based on Frederic Piguet calibres 70 and 71), measuring only 2.4 mm and highlighting silicon pallets and free-sprung[215] balance spring. It carries a power reserve of forty-five hours.

Grand feu enamel dials are fired in a kiln heated at about 830°C (1,526°F)—justifying the technique's name, which translates from French as "great fire."

Enamel powder is scattered onto a base dial, fusing together in a kiln at exceptionally high temperatures. Extreme care is necessary to ensure that the enamel layer does not distort or crack. Application of successive layers between the six to eight firings produces the uniquely iridescent white sheen.

As is the case with other Breguet timepieces, there is a secret signature on the dial. Engraved in an exceptionally sharp and tiny script right into the enamel between the cannon pinion and the three o'clock marker, the word "Breguet" provides an added level of anticounterfeiting security.

Between seven o'clock and eight o'clock on the flange beneath a decorative minute track, the words "Swiss Emaille Grand Feu" appear in black lacquer. This ornamental circle includes vintage star shapes that have been in the Breguet catalog of symbols for centuries. The printed numerals appear flawless.

Perhaps the most beautiful feature is the off-center sunken subsidiary seconds, beautifully integrated into the (exceptionally difficult to make) one-piece dial. By ensuring a single dial structure, Breguet artisans facilitate the printed number "6" spilling over into the delicately concave recess of the subseconds indentation.

The beauty of enamel is not just in its deeply lustrous patina, but also in the fact that its opalescent shine will last as long as the watch itself, without any fear of discoloration. Since enamel is glass, not paint, it does not fade when exposed to ultraviolet light.

214. Enamel, also called vitreous enamel or porcelain enamel, is a material made by fusing powdered glass onto a surface material. After the glass melts and flows across the surface material, it hardens to a smooth, durable vitreous (meaning glass—from Latin vitreum) finish.

215. A balance spring without a regulator, commonly called a "free-sprung" balance, generally provides superior accuracy to its regulated brethren. To adjust the rate, instead of working on the spring length (which is how a regulator works), watchmakers modify the inertia of the balance wheel, moving screws or weights positioned on the balance.
Unencumbered by regulator pins, the balance spring exhibits a constant length and oscillates with perfect concentricity. It is, however, more difficult to service, since it requires symmetrical adjustment to maintain poise.

Breguet
SWISS EMAILLE GRAND FEU

MILITARY

Synchronizing battle maneuvers without signaling proved popular toward the end of the nineteenth century. It was then that military men first wore wristwatches. The Garstin Company of London patented a "Watch Wristlet" design in 1893, but these timepieces may have been extant from the 1880s. Officers in the British army began using wristwatches during colonial military campaigns in the 1880s. During the First Boer War, coordinating synchronized troop movements and attacks against the highly mobile Boer insurgents was of utmost importance, and among the officer class, the subsequent use of wristwatches was widespread. Mappin & Webb began producing their successful "campaign watch" for soldiers during the Sudanese campaign in 1898 and accelerated production for the Second Boer War a few years later. Girard-Perregaux and other Swiss watchmakers began supplying German naval officers with wristwatches in about 1880.

These early models were essentially standard pocket watches fitted to a leather strap, but by the early twentieth century, manufacturers began producing purpose-built wristwatches. The Swiss company Dimier Frères & Cie patented a wristwatch design with the now-standard wire lugs in 1903. Hans Wilsdorf moved to London in 1905 and set up his own business, Wilsdorf & Davis, with his brother-in-law Alfred Davis, providing high-quality timepieces at affordable prices; the company later changed its name to Rolex. Wilsdorf was an early convert to the wristwatch and contracted the Swiss firm Aegler to produce a line of wristwatches.

The impact of the First World War dramatically shifted public perceptions on the propriety of a man's wristwatch and opened up a mass market in the postwar era. The "creeping barrage artillery tactic" developed during the war required precise synchronization between the artillery gunners and the infantry advancing behind the barrage. Service watches produced during the war were especially useful for the rigors of trench warfare, sporting luminous dials and unbreakable glass. The British War Department began issuing wristwatches to combatants from 1917. By the end of the war, almost all enlisted men wore a wristwatch, and after demobilization the fashion soon caught on.

When Casio introduced the G-Shock (G is for gravitational) in 1983, it quickly garnered the attention of military personnel throughout the world. The watch, intended primarily for sports, military, and outdoors-oriented activities, resists mechanical shock and vibration. It is now also available in all-metal varieties.

The US Department of Defense (USDoD) issues military equipment standards known informally as MilSpecs. Many watch manufacturers make timepieces that conform to the USDoD's specifications. Function determines form for the military watch. These watches provide optimum performance on the battlefield. The sandblasted matte finish on the typical military watch is not accidental; an unintentional glare from a shiny watch surface may give away a soldier's position. Sandblasted finishing also facilitates corrosion resistance. The military timepiece is minimalist in design; nothing superfluous appears on the dial or the case. There is usually a twenty-four-hour marking on the watch to aid military time protocols. In addition, luminescent markings are required, usually triangular shaped and placed at each hour. There are three main types of military watches: field watch, pilot's watch, and dive watch. Look for descriptions of the latter two in other parts of this book. The field watch or general service watch caters primarily to infantry units.

MIL-PRF-46374G is a performance specification document issued on November 12, 1999, by the US Department of Defense. This document elucidates such characteristics as shock protection, working temperature range, water resistance, antimagnetism, altitude, perspiration, and readability in total darkness.

AVIATION/PILOT'S WATCH

Louis Cartier made the first aviation timepiece for a friend of his, Brazilian aviation pioneer Alberto Santos-Dumont. In 1904, Santos-Dumont complained to Cartier about the difficulty of checking his pocket watch to time his performance during flight. Santos-Dumont then asked Cartier to come up with an alternative that would allow him to keep both hands on the controls. The result was the first Cartier men's wristwatch, with a leather band and a small buckle.

Santos-Dumont always flew with his Cartier wristwatch. Today, Cartier markets wristwatches and sunglasses named after Santos-Dumont. In 1909, Louis Bleriot flew over the English Channel wearing a Zenith, which he later publicly applauded for its reliability.

Modern pilot's watches generally feature a stop-seconds function, low-light visibility, precise timekeeping in temperature extremes, and a strap designed to be worn over a flight suit. Aviation watches feature an arrow or triangle at the twelve o'clock position. This marker orients the pilot in low-visibility situations.

Cartier Santos Dumont Reference 1575 (ca. 1994). Limited edition of ninety examples in platinum, with magnificent salmon dial commemorating the ninetieth anniversary of the first Santos from 1904.

As with the diving watch, the aviation watch has largely become outmoded, with numerous technological improvements and a vast array of electronic instruments present on today's aircraft. These timepieces, however, serve as a valuable backup in cases where the onboard instrumentation has failed.

The most famous instrument aviation watch is arguably the Breitling Navitimer, still in production. Breitling also produces a wristwatch specifically conceived to aid in search-and-rescue operations, should the wearer be stranded. It goes by the name "Breitling Emergency" and is the first wristwatch equipped with a miniaturized dual-frequency distress beacon (operating on the 121.5 MHz international air distress frequency), designed to issue alerts and guide search and rescue missions up to 100 miles away.

Breitling Navitimer 01: The definitive aviator reference first introduced in 1952 (although the Chronomat, released ten years prior, introduced the slide rule).

The slide rule on the flange operates by moving the bidirectional bezel to perform numerous calculations using logarithmic scales—multiplication, division, speed and distance, and currency and temperature conversions.

For in-flight use, the slide rule can calculate fuel consumption, rate of climb or descent, and distance of climb or descent. The Navitimer also features a chronograph and seventy-hour power reserve.

MPH
KM
BASE
1000 M
NAUT.
STAT.
BREITLING
1884
CHRONOMETRE
NAVITIMER
SWISS
MADE

FITNESS

Since the late 1970s, there have been a number of fitness crazes. With each fad comes the concomitant wristwatch. In the '80s, it was a Casio offering digital timekeeping in a sturdy black polycarbonate case. Modern fitness watches are purpose-built devices designed to track everyday activities and improve one's overall health. Fitbit devices adorn the wrists of millions of fitness fanatics worldwide. These devices provide timekeeping as an important but secondary function. The primary function of these modern activity trackers is to monitor steps climbed, distance run, calories burned, heart rate, and sleep, among other things.

The Apple Watch is undoubtedly the definitive smartwatch, also offering cell phone functionality independent of the iPhone. Its fitness functions continue to improve with each new iteration. The Apple Watch Series 5, shown here (although superseded by the Apple Watch Series 6 in September 2020), has an always-on "Retina" display and offers a host of health-related features, including built-in ECG functionality. When the user holds the crown with the opposite hand, the watch reads the heart's electrical signals and provides notifications via the Heart Rate app to alert the wearer regarding any heart-related irregularities.

High-intensity interval training, running, cycling, yoga, and swimming are among the many activities the watch can track.

The Activity screen shown here displays three "rings," with each one representing a percentage of goal completion for three distinct measures. Starting with the blue inner ring, there is an initial "stand" goal of moving at least one minute during twelve different hours in the day. The green ring is the "exercise" ring, which can close and be exceeded (the moving hemispheric color separation indicates the current position—twelve o'clock is 100 percent of goal) with at least thirty minutes of activity at brisk walking pace or faster. The red ring is the "move" ring and shows percentage of calories burned.

The geniuses within the Apple Design department even incorporated the ring names and intentions into the arrow design. The arrow pointing to the right of the red "move" ring indicates, with its singular arrow, basic movement. The twin arrows pointing to the right on the green ring suggest with their double pointedness the get-moving sign for "exercise," and the up arrow simply stands (yes, pun intended) for standing up.

Activity
11:49

DRESS

The dress watch is the most sophisticated and stylish of watches, complementing a business suit or formal attire. This watch, typically of stainless steel or precious metal, is slim, elegant, and at home at the opera, ballet, symphony, fine restaurant, banquet, state dinner, or other regal celebration or event. Most collectors or connoisseurs have at least one dress watch in their collection, and often it serves as the pinnacle of the collection.

Breguet Classique 5177BA/12/9V6 self-winding dress watch in yellow gold, with "guilloche" dial produced using a rose engine and 0.10 mm precision.

Look closely at either side of the Roman XII and you will see "Breguet" gently engraved into the silvered 18K gold dial. Abraham-Louis Breguet introduced the crescent-tipped hands, famously called "Breguet" or pomme hands, in 1783. The hands are thermally induced blued steel.

XII
XI
BREGUET
X
IX
VIII

SPORTS

The sports watch may include watches from other categories, including the dive watch, aviation watch, military watch, and fitness watch. Typically, the sports watch is an all-purpose timepiece with a multitude of functions, including chronograph and alarm. It may also include other useful features such as GPS, altimeter, barometer, compass, multiple time zones, and countdown timer. It will likely be at home in a moisture-rich, hot or cold, high-altitude, or high-pressure environment.

Characteristically, sports watches are water and shock resistant and are commonly purpose-built for a particular sport or activity.

Garmin's fēnix 6X Pro Solar Edition. Considered by many to be the best all-around sports watch of any kind. The uniquely spelled name (fēnix) with the "macron" diacritical mark is a registered trademark. The macron (meaning "long" in ancient Greek) elongates the hard e-sound so that the model name is pronounced just like the ancient Greek mythological bird, the phoenix. (The myth goes that the phoenix is reborn in an ever-repeating cycle from the residual ashes of an incineration of its own making hundreds of years after its last reincarnation. Garmin is up to its sixth edition of the fēnix and doesn't seem to be slowing down, so the moniker is well chosen.)

The Garmin fēnix 6X Pro Solar Edition comes in a variety of case finishes and straps; the titanium model appears here. (Ref. 010-02157-23, current MSRP: $1,149.) The 1.4-inch-diameter (35.56 mm) display features 78,400 pixels, visible even in sunlight. The watch is water resistant to 100 meters. Battery life is up to twenty-four days, with built-in solar charging.

Modern sports watches, especially those powered by powerful CPUs, have myriad features, including, of course, basic clock features. In addition to GPS-synchronized time and date, alarm, timer, stopwatch, and sunrise/sunset times, the exceptionally powerful and capable smartwatch also includes heart rate and respiration rate monitoring; pulse oximeter blood oxygen saturation; stress tracking; relaxation reminders; sleep monitoring; calendar; weather; music storage; smartphone compatibility; step counter; calories burned; floors climbed; distance traveled; intensity minutes; cardio, strength, yoga, and Pilates workouts; training, planning, and analysis features; running features, including stride length, cadence, and even lactate threshold (with compatible accessories); golfing features, including a digital scorecard and a full vector map of thousands of (downloadable) golf courses; outdoor activities, including preloaded ski resort maps and elevation profiles; cycling features; and swimming features, including stroke type detection and pool swim metrics such as pace and stroke count.

ALTIMETER
5545
8600
5464
Last 4 Hours
GARMIN
LIGHT
UP·MENU
DOWN
55
05
45
15
35
25

SKELETON

The skeleton watch is to watchmaking what the visible engine bay is to the automotive supercar. It provides a complex view of the extraordinary engine that lies within. Typically, the skeleton watch features a partially or even fully removed dial that normally obscures the view of the gorgeous movement below. In some cases, the movement itself is "skeletonized" by trimming away nonessential metal on the bridge, plates, wheel train, or other mechanical parts. Often, the remaining structure is beautified through engraving, beveling/chamfering, or specialized decoration. Every time you look at your wrist, you see the microengineered machine that powers your glorious timepiece!

TAG Heuer's Calibre Heuer 01 (Ref. CAR2A1T.FT6052, MSRP: $5,450) with in-house movement is a good example of a timepiece that serves as both a sports watch and a skeleton watch. It features a perforated blue rubber strap, 100 meters of water resistance, a polished steel and black PVD case, and a sapphire crystal front and back.

TACHYMETRE
TAG HEUER
CARRERA HEUER 01
CHRONOGRAPH
AUTOMATIC
SWISS MADE

"Time plays like an accordion in the way it can stretch out and compress itself in a thousand melodic ways. Months on end may pass blindingly in a quick series of chords, open-shut, together-apart; and then a single melancholy week may seem like a year's pining, one long unfolding note."

–Julia Glass, *Three Junes*

ASTRONOMICAL

The movement of Earth relative to the celestial bodies formed the basis of historical time measurement. Since our concept of time intimately interweaves with the movement of these cosmic entities, the fascination with astronomical complications in watches has grown ever more acute. Watchmakers continually seek to include an expanding array of complications into timepieces, and the wide selection of astronomical complications makes for excellent candidates. The more notable astronomical functions include equation of time, star maps, sidereal time, and sunrise and sunset.

The Patek Philippe Celestial (Ref. 6102P-001, ca. 2015, MSRP: $329,600) is a towering achievement of astronomical precision. The background consists of a blue sapphire-crystal disc tracking the orbital position of the moon. Another small sapphire-crystal disc displays the moon phases.

On top of these is a transparent sapphire-crystal disc that depicts an authentically detailed sky chart surrounding the Milky Way. **The area within the white elliptical border depicts the sky as it appears from Geneva and locales at similar latitudes**. The astronomical features in this timepiece are so precise that deviation for the lunar day is a mere 18.385 seconds **per year**. For a sidereal day, the deviation is 32.139 seconds per year. The moon phase is accurate to within 6.51 seconds per lunation.

PATEK PHILIPPE GENEVE
S
W
E
N

EQUATION OF TIME

The equation of time describes the discrepancy between two kinds of solar time. "Equation" is in reference to the medieval interpretation of "reconciling a difference." The two times that differ are the apparent solar time, which directly tracks the motion of the sun, and mean solar time, which tracks a theoretical average sun with noons twenty-four hours apart. Apparent or true solar time, as indicated by a sundial, represents the current position of the sun. Mean solar time, for the same place, would be the time indicated by a steady clock set so that over the year its differences from apparent solar time would resolve to zero.

Apparent time (and the sundial) can be ahead (fast) by as much as sixteen minutes and thirty-three seconds (around November 3) or behind (slow) by as much as fourteen minutes and six seconds (around February 12). A watch equipped with an equation-of-time function typically utilizes a sweeping hand on a subsidiary dial graduated along an arc registering −15 to +15 minutes. To calculate the true time, a little mental arithmetic is required.

Martin Braun's EOS Boreas (Ref. MB16142S, ca. 2003), showing sunrise (*left*) / sunset (*right*) indicators with visible cams (elliptical-shaped rotating components that create a prescribed motion) at six o'clock and equation of time indicator below the date in the center of the dial. Powered by Calibre 2892-A2 with MAB 88 module (twenty-five jewels). Custom-calibrated indicators represent each owner's particular geographical location.

MARTIN BRAUN
EOS
MADE IN GERMANY

STAR MAPS

In recent years, premium watch brands have offered custom dials for very exclusive timepieces to well-heeled individuals. These custom dials feature the night sky as it appears from the owner's home city location. The most famous and popular modern example is arguably Patek Philippe's Sky Moon Tourbillon. In addition to featuring a host of other features, including sidereal time (see below), it also exhibits on its reverse side a finely detailed sky chart (star map) of the night sky as it currently appears above the owner's city, including the richly detailed band of the Milky Way (see caption on page 198).

SIDEREAL TIME

This method of timekeeping involves following a star (other than our sun) over a mean sidereal day of twenty-three hours, fifty-six minutes, and 4.09 seconds. It is this amount of time required for Earth to rotate once on its axis. The term takes its name from the Latin word *sidus*, meaning star.

MOON PHASES

The traditional moon phase consists of a fifty-nine-toothed wheel measuring a 29.5-day month. However, the moon's rotation around the earth (the synodic month) varies but averages a duration of twenty-nine days, twelve hours, forty-four minutes, and 2.8016 seconds (see earlier chapter on the measurement of time). To measure duration more accurately, most moon phase timepieces are equipped with a 135-toothed wheel, improving precision to one day in approximately 122 years. In 2014, Andreas Strehler introduced the Sauterelle à Lune Perpétuelle, a moon phase watch that is accurate to one day in a barely believable 2,060,757 years.

Frederique Constant's Ref. FC-715MC4H4 (ca. 2017) features a gorgeous hunter case back and in-house calibre FC-715 (ca. 2014) with central seconds, date, and moon phase (set via the crown).

FREDERIQUE CONSTANT
GENEVE
SWISS
MADE

REPEATERS AND SONNERIES

The minute repeater chimes the time down to the minute, using separate tones for hours, quarter hours, and minutes. These exceptionally complicated timepieces originated in a time before the adoption of widespread artificial illumination. Then, reading off the time at night on one's clock or watch—when the only illumination was from moonlight or starlight—was difficult, if not impossible. Initially, bells mounted in the back of a watchcase struck the tones. However, the development of wire gongs during the nineteenth century reduced the space necessary to house the components.

There are other types of repeaters, including the quarter repeater, the decimal repeater, the half-quarter repeater, and the five-minute repeater—all of which sound out the number of intervals that have elapsed since the start of the hour.

Different to the repeater but related in terms of ability to sound out the time are sonnerie watches. On the hour and the quarter hour in passing, they chime the time on tiny gongs. Importantly, they do not require on-demand activation from the wearer, as is the case with repeaters. Grand sonneries (sonnerie means "alarm" or "strike") feature a quarter striking mechanism combined with a repeater. They are so difficult to manufacture that only recently (1992) was a wristwatch fitted with a grand sonnerie, when master watchmaker Philippe Dufour incorporated it into his magnum opus: "Philippe Dufour Grande and Petite Sonnerie." Petite sonneries merely strike the hours and quarter hours with no repeater function. In addition to grande sonnerie and petite sonnerie, some watches offer other "striking" mechanisms. The Vacheron Constantin Reference 57260 features the Westminster carillon chiming, utilizing five gongs and five hammers. It is the most complicated watch striking mechanism yet devised. The Westminster carillon is the chime of the world-famous clock of the British Parliament building in London, a.k.a. "Big Ben."

IWC's Il Destriero Scafusia (the Warhorse of Schaffhausen), Ref. 1868 (42.2 mm diameter, 18 mm thickness), released to coincide with the 125th anniversary of the storied brand. Following a high-luxury Swiss watchmaking renaissance and not wishing to be outclassed by Patek Philippe (with its Calibre 89 pocket watch), Blancpain (with its model 1735), and other modern complications, in 1993 this IWC became **the world's most complicated mechanical wristwatch**. Limited to just 125 examples, the majority left the factory encased in yellow gold, with additional examples in rose gold and platinum. The Grande Complication combined a split-second chronograph, flying tourbillon, perpetual calendar (with four-digit year display), and minute repeater (with repeating hours, quarters, and minutes). Remarkably, the manual-wind, seventy-six-jeweled calibre 18680 utilizes a Valjoux 7750 as its base, rendered completely unrecognizable in the finished timepiece. The onyx-adorned crown activates the split-second function. Actuating the slide on the left case band chimes the hours, quarters, and minutes on demand.

IWC
SCHAFFHAUSEN
TOURBILLON
SUN MON TUE WED THU FRI SAT
SEP OCT NOV DEC JAN FEB MAR APR MAY JUN JUL AUG
2019
SWISS
MADE

TOURBILLONS

The tourbillon, from the French word for whirlwind, mounts the escapement and balance wheel in a rotating cage, negating the effect of gravity when the timepiece (and thus the escapement) rests in a certain position.

First developed around 1795, the tourbillon was patented by French Swiss watchmaker Abraham-Louis Breguet on June 26, 1801. By continuously rotating the entire balance wheel / escapement assembly usually one revolution per minute, positional errors average out.

Gravity directly affects the most-delicate parts of the escapement, specifically the pallet fork, balance wheel, and hairspring. The hairspring, which functions as the timing regulator for the escapement, is extremely sensitive, especially to magnetism, shocks, temperature, pinning point, terminal curve, and heavy points on the balance wheel. Many inventions exist to counteract these problems. Newly developed materials ameliorate temperature and magnetism problems. Shocks have much less effect today than at Breguet's time, thanks to stronger and more-resilient materials. Gravity comes into play on the remaining effects.

A tourbillon most often makes one complete revolution per minute (although thirty-second tourbillons are available from several luxury watch brands). This improves timekeeping because—even if a watch is immobile in a random vertical position—the tourbillon turns the escapement around its own axis, effectively ironing out the effects of gravity by turning the balance through all possible vertical positions during its rotation. A normal tourbillon has no real effect in horizontal positions, since in this case the balance is horizontal and not affected by gravity as it turns.

Because pocket watches were usually carried in one's waistcoat pocket, and thus aligned vertically, Breguet designed the tourbillon to substantially reduce gravity's impact. However, unlike a pocket watch, the attitude (positioning) of a wristwatch changes frequently from vertical to horizontal depending on what its wearer is doing. The effect of a tourbillon in a wristwatch is therefore insignificant compared to the change in rate resulting from changes from vertical to horizontal and vice versa.

In modern mechanical watch designs, a tourbillon is unnecessary to produce a highly accurate timepiece. Nevertheless, the tourbillon is one of the most valued features of collectors' watches, and premium timepieces often feature a tourbillon as an exemplar of a brand's watchmaking acumen.

A close-up view of Breguet's Messidor Tourbillon (5335BR/42/9W6) as seen from the case back. The tourbillon itself appears to float in the air, but sandwiched sapphire crystals actually hold it in place.

The "Messidor" name is in tribute to Abraham-Louis Breguet, the tourbillon's inventor and namesake of the Breguet brand (now a subsidiary of the Swatch Group). The invention was patented on 7 Messidor year IX (as inscribed on the tourbillon's circumference) per the French Republican calendar, which was used by the French government for about twelve years from 1793 to 1805.

MESSIDOR AN

CHRONOGRAPHS

Together with calendar complications, the chronograph is perhaps the most sought-after complication for collectors. The term "chronograph" comes from the Greek words for time and writing, *chronos* and *graph*. Essentially, chronographs record the timing of an event at the bidding of the wearer. Chronograph timing is independent of the basic time-telling function, although most chronographs utilize the same gearing and escapement. A typical chronograph has an independent-sweep second hand that can be started, stopped, and returned to zero by using either one or two buttons. Most chronographs, especially those found on mechanical timepieces, utilize a start/stop pusher at two o'clock and a reset pusher at four o'clock. There are many different kinds of chronographs, differing in terms of complexity of manufacture, basic layout, and overall function.

If it takes **twenty seconds** to cover 1 mile, the read-off on the bezel is **180 mph** (see red numerals on bezel).

If it takes **25.71** seconds to cover 1 mile, the read-off on the bezel is **140 mph** (see blue numerals on bezel).

The formula for velocity is distance divided by time—for example, 1/25.71 = 0.03889 . . . (miles per second) × 60 seconds = 2.333722 . . . (miles per minute) × 60 minutes = 140 (miles per hour). Similarly, if you traverse a kilometer in twenty seconds, you are moving at 180 kph (or 140 kph in 25.71 seconds). One can easily intuit that if 1 mile or kilometer took sixty seconds to complete, the velocity would be 60 mph or 60 kph.

The bezel on a chronograph often features a measurement scale referred to as a tachymeter, particularly useful in computing average velocity over a specified distance. The Rolex Daytona illustrated here displays "units per hour" engraved around the black ceramic bezel and demarcated at precisely measured intervals. Other tachymeters exhibit similar markings (although not necessarily with the same velocities), which sometimes appear on the flange (inside the dial) instead of on the bezel. Before digital instrumentation, race car drivers and other rapid travelers used the bezel markers to measure average velocity (speed) over 1 mile or 1 kilometer. Starting the chronograph at launch (zero distance), the pusher at two o'clock is pressed once again at the end of the measured distance, and the speed is read off from the bezel.

A modern classic, the Breitling Avenger II chronograph. Style: A13381111B2A1 with "Volcano Black" dial displaying classic Breitling stenciled numerals.

Considered an elite tool watch, the Avenger II chronograph features water resistance of 300 meters (1,000 feet) and a self-winding chronograph; like all Breitling calibres, it is chronometer certified.

The three-register layout at 12, 6, and 9 features a thirty-minute counter, twelve-hour counter, and running seconds, respectively. The central red-tipped second hand indicates chronograph timing. The raised minute scale at the cardinal points of the bezel is also signature Breitling. Known as "riders," these "tabs" provide grip and additional protection to the crystal and bezel.

BREITLING
1884
CHRONOMETRE
CERTIFIE
AUTOMATIC
SWISS MADE

FLYBACK

To assist pilots, the flyback chronograph eliminates the time loss while performing the stop/reset function. Every second counts during flight, especially in an age prior to electronic navigational equipment. A loss of a second or more while engaging the stop button, the reset button, and then the start button (again) to begin timing a new event would result in significant navigational inaccuracies. The flyback overcomes the issue of multiple button pushes by allowing the chronograph hand to reset automatically to zero and begin recording again with only a singular button push—usually the same button at 4 used to reset the chronograph.

Blancpain's 38 mm Leman Flyback Chronograph (Ref. 2185F-1130-71, ca. 2000–2009). Upon its debut, this iconic column-wheel, vertical-clutch flyback chronograph became an instant success. It features **one of the finest chronograph movements of its time.**

BLANCPAIN
Flyback
SWISS MADE

RATTRAPANTE

Also known as a double chronograph or split-seconds chronograph, the rattrapante, from the French word *rattraper* (to catch up), is considered one of the most difficult complications to reliably manufacture. Watch aficionados especially cherish vintage examples. This type of chronograph features two chronograph hands superimposed one above the other. An additional push button is usually employed to activate the second chronograph hand and is (classically) either embedded in the crown or located at the ten o'clock position on the case. Initial operation follows the standard method of starting the chronograph. The first push releases both hands (operating in tandem, one above the other). While one hand continues registering the time, repeated stoppage of the other hand provides read-off of the "interim" time. After the interim time is recorded, the rattrapante push piece is pressed to "catch up" or synchronize the split-second hand once again with the main chronograph second hand. This type of complication excels at timing events involving laps completed by the same individual or in timing two individuals competing in the same event.

The manual winding Breguet Rattrapante (Ref. 5947BB, ca. 2006, MSRP: $56,900) in 18K white gold. The inset crown button controls split-seconds functionality.

The lightly engraved Breguet "secret signature" (to thwart counterfeiting) is located on either side of the Roman numeral XII. The dial features numerous styles of rose-engine-turned guilloche (see section on "Decorative Watchmaking Techniques").

BREGUET
RATTRAPANTE
SWISS GUILLOCHE MAIN

FOUDROYANTE

The foudroyante (food-ro-yahnt)—from the French *foudre*, meaning lightning—features either a central hand or subdial hand that enables fractional-second measurement. The foudroyante typically measures quarters, fifths, sixths, eighths, or tenths of a second. This feature is normally part of the chronograph, but it is not required to be included with the chronograph complication.

The flange of the Zenith El Primero Foudroyante one-tenth of a second chronograph (Ref. 03.2041.4052/69.C496, ca. 2010), calibrated for one-tenth of a second display.

Zenith had to optimize the movement by making silicon wheels to ensure that the movement parts were as light as possible, to accommodate the relatively extreme accelerative forces and torque of the red jumping second hand.

ZENITH
EL PRIMERO
1/10 OF A SECOND

ALARMS

Quartz watches frequently offer an alarm function. In mechanical watches, the complication first appeared in 1914 but gained popularity only in the 1940s and '50s with the introduction of the Vulcain Cricket (1947), so named for its cricket-like alarm sound. Until then, no watchmaker was successful in developing a wristwatch alarm that was powerful enough to awaken its wearer.

The Vulcain Cricket comprised two barrels, one for the movement and the other for the alarm. Several presidents wore the Vulcain Cricket, starting with Harry S. Truman in 1953, resulting in the brand's slogan "the Presidents' Watch." These watches offer an additional "alarm" hand that can be set for a predetermined time.

Vulcain and wristwatch alarms are almost synonymous—the company being the inventor of the alarm complication for wristwatches.

Shown here is the skeletonized Vulcain Anniversary Heart (Ref. 180128.256, MSRP: $7,500), commemorating the brand's 150th anniversary and limited to 150 examples each in three assorted dial colors. The red-tipped arrow-hand adjusts the alarm in ten-minute increments.

VULCAIN
SWISS MADE

MULTIPLE TIME ZONES AND GMTS

Modern business travelers often have to juggle knowing the time in two or more time zones. Multiple time zone, Worldtimer, GMT, and UTC are interchangeable terms referencing the ability to tell the time in several time zones with ease.

A BRIEF HISTORY OF TIME ZONES

Without the twenty-four time zones radiating from Greenwich in England to the rest of the world, trade and communication would certainly be more difficult. Nineteenth-century globalization owes a great deal to standardization of international time. Even as late as the 1940s, some circles still debated the use of standardized clocks.

Let us step back a bit in history. In 1875, the United States railway system recognized seventy-five different "local" times,[216] three of which were in Chicago alone.

In 1883, at a meeting of the American Metrological Society,[217] it was resolved that "a committee of the Society be appointed to draft a memorandum to be addressed to the managing officers of the railways of the United States and Canada . . . urging the speedy adoption by the railway companies of the hour system of time standards based upon the 75, 90 and other hour meridians."

Just how did they arrive at this resolution? Within the twelve preceding months, German professor W. Förster, director of the Astronomical Observatory (at Berlin), in considering the establishment of a prime meridian[218] (with reference to German normal time), advocated for the "selection of the anti-meridian of Greenwich for a common zero of longitude and time."[219] European, British, and Russian scientists similarly concurred, with principal focus on the establishment of a prime meridian somewhere in the world and secondary focus dealing with the specific location of said meridian.

On August 3, 1882, "the Senate and House of Representatives passed a joint resolution authorizing the President of the United States to call an International Conference to fix and recommend for universal adoption a common prime meridian, to be used in the reckoning of the longitude and the regulation of time throughout the world."[220]

Getting to this point was arduous. Not only were different cities within the same country operating on competing times, but as recently as 1918, Russia's calendar indicated a *thirteen-day discrepancy* with western Europe (refer to the section on the Gregorian calendar for an explanation). Local and even international logistics were being constrained by a hodgepodge of times. An inability to synchronize schedules stymied the movement of troops and goods throughout the Eurasian landmass. Indian society, at the time under British rule, found the imposition of another imperial directive repellant. Other societies similarly pushed back.

Carl F. Bucherer's Patravi TravelTec GMT displaying three time zones simultaneously. Ref: 00.10620.03.33.01, ca. 2010, current MSRP: $41,800.

216. Ian P. Beacock, "A Brief History of (Modern) Time," *The Atlantic*, December 22, 2015, https://www.theatlantic.com/technology/archive/2015/12/the-creation-of-modern-time/421419/.

217. *Proceedings of the American Metrological Society* 4 (1883).

218. The earth's zero of longitude, which by modern convention passes through Greenwich, England.

219. Ibid.

220. Ibid.

East
West
CARL F. BUCHERER
TravelTec
GMT
SWISS
MADE

In the United Kingdom, the Isle of Man was the first to adopt Greenwich Mean Time (GMT), in 1883. In the same year, at precisely noon on November 18, North American railroads switched to a new standard time system for rail operations, which they called Standard Railway Time (SRT).[221] Four time zones were adopted; namely, Eastern, Central, Mountain, and Pacific. Nationwide adoption of these standards quickly followed, and Daylight Saving Time took effect in 1918 with Congress's passing of the Standard Time Act.[222]

Technically speaking, a GMT watch would indicate the difference between the wearer's home time—say, New York (GMT −5, or GMT −4 in summer)—and the time at the Royal Observatory in Greenwich, United Kingdom.

Since 1972, Universal Time Coordinated (UTC) has replaced GMT as the standard designation for referencing time zones. UTC is the International Atomic Time and accounts for leap seconds added periodically to compensate for the earth's minor rotational deceleration. In the strictest sense, UTC is a timekeeping system and not a time zone at all. Nevertheless, for casual use, GMT and UTC are equivalent.

Time zones around the world are expressed using positive or negative offsets from UTC. The westernmost time zone uses UTC−12, being twelve hours behind UTC; the easternmost time zone, theoretically, uses UTC+12, being twelve hours ahead of UTC.

Numerous wristwatches currently provide multiple time zone abilities. Carl F. Bucherer manufactures a triple-time-zone watch, the Patravi TravelTec, which elegantly presents three time zones utilizing concentric twenty-four-hour displays.

In 2011, Vogard released the Chronozoner, which allowed wearers to adjust the time instantaneously to a preset time zone via the use of a lever, which unlocked the city ring bezel. Turning the city ring to the desired city (at twelve o'clock) displayed the current time in that city.

Ulysse Nardin's two-button system for advancing and retreating the hour on the basis of the new location also includes the ability to keep the current date in synchronization.

Vacheron Constantin features a rotatable Lambert projection world map on some of its world timers, around which a translucent lacquered disc bears the city names. In addition to the cities disc, a day/night tinted sapphire crystal overlays the area of the world currently lit by the sun.

CALENDAR WATCHES

From the basic calendar displaying just the day of the month to a perpetual calendar accurately displaying the Gregorian date format (at least until the upcoming centenary year—see earlier discussion on calendars), the calendar function has always been one of the most sought-after complications, not least because of its practicality.

The perpetual calendar remains the province of only the most-talented master watchmakers. This highly sophisticated mechanical memory integrates into the watch in a module that reproduces the calendar system and automatically accounts for months with thirty or thirty-one days and twenty-nine days in February of each leap year.

The A. Lange & Söhne Datograph Perpetual (Ref. 410.032 E) featuring calibre L952.1 with 556 parts (223 components comprise the calendar mechanism). At 41 mm, it features a manually wound thirty-six-hour power reserve oscillating at 18,000 vph. The silver dial with applied rose gold indices features rose gold alpha hands, **perpetual calendar** with month subdial at four o'clock (with inset leap-year subdial at 4:30), day-of-week subdial at eight o'clock, a.m./p.m. indicator arrow at nine o'clock, moon phase at six o'clock, flyback chronograph with tachymeter scale on flange, big date at twelve o'clock, and sapphire crystals front and back. Approximate price at launch: $136,600.

221. "Today in History—November 18," Library of Congress, https://www.loc.gov/item/today-in-history/november-18/.

222. There is a great deal more to say about this topic; I would recommend *The Global Transformation of Time, 1870–1950* by Vanessa Ogle (Cambridge, MA: Harvard University Press, 2015) for further discussion.

A. LANGE & SÖHNE
DATOGRAPH
PERPETUAL
GLASHÜTTE I/SA
MADE IN
GERMANY
BASE 1000 METERS

The perpetual calendar requires no adjustment until 2100. However, for those who do not want to make even the single-day adjustment for February 28, 2100 (since there is no February 29, 2100—to be a leap year, a century year must be divisible by 400—see earlier discussion on leap years), the Patek Philippe Calibre 89 is the watch for you. This pocket watch features a patented star wheel that goes around once every century. You may need deep pockets, not only from a practical perspective (it weighs about 1.1 kg / 2.4 lbs. and has a diameter of 88.2 mm), but financially too. The Calibre 89, a set of four pocket watches, each with 1,728 parts and thirty-three complications, and taking nine years to manufacture, was released by Patek Philippe in 1989. One of these extraordinary timepieces sold for over $5 million at auction in Geneva in 2009.

The most common date complications are the following:

- the regular date window aperture (day of month, usually located at three o'clock, 4:30, six o'clock, or nine o'clock)
- big date (day of month displayed typically with two large date wheels, usually at twelve o'clock)
- pointer date (day of month on perimeter of dial)
- subsidiary date (date displayed on a subsidiary dial)
- day-date (a combination of the day of the week and day of the month shown either next to each other or in different parts of the dial)
- the triple calendar (also called a complete calendar featuring the name of the month in addition to the day of the week and date)
- the perpetual calendar (day, date, month, and leap year indications)
- the annual calendar (same as the perpetual calendar but without leap year functionality, requiring annual February 29 adjustment)
- the equation of time (refer to the astronomical section above)

RETROGRADES AND JUMPING

The retrograde watch displays a function with a linear jumping hand (e.g., a timepiece may display the day of the week in a semicircular fashion). At midnight (or nearly midnight) on Sunday, the retrograde day-of-the-week hand will nearly instantaneously swing back to Monday. A jumping feature (typically used for hours or minutes) utilizes numerals that "jump" to display the time. Instead of typical hands representing the hours and minutes, at the completion of a minute or hour, digits on a disc rotate instantaneously to the next position.

Vacheron Constantin Patrimony
Ref. 86020/000G-9508 (last known MSRP: $46,200, ca. 2010). Case in 18K white gold. In addition to the standard hour and minute hands, the classically styled timepiece features two retrograde hands: one for the day of the month, the other for the day of the week. When the indicator reaches the end of the hemisphere, it jumps back to the beginning.

VACHERON CONSTANTIN
GENEVE
MONDAY
TUESDAY
WEDNESDAY
THU
FRIDAY
SATURDAY
SUNDAY
SWISS
MADE

TAKING CARE OF YOUR WATCH

"Time and distance
Both illusions of the mind
No beginnings, no ends"

–Alex Z. Moores

Taking care of a mechanical timepiece is relatively straightforward, but unintended magnetization of the balance and water ingress are two of the most common issues that novice watch owners should know about and diligently seek to avoid. One should not expose the watch to strong magnetic fields, which are all around us in assorted appliances but especially in common household items such as speakers and vacuum cleaners. A magnetized balance will not keep good time.

Do not knock or drop the watch—this will unsettle the parts inside. While many modern timepieces have one or more shock absorption devices, they are not always sufficient to protect against damage.

Always set the date between 3 a.m. and 9 p.m., or, inversely, never set the date between 9 p.m. and 3 a.m. Date wheels are fussy; they do not appreciate manual adjustment either side of midnight, and irreversible damage may result. Some modern watchmakers have devised date change systems to ameliorate this concern, but it is good practice never to set the date either side of midnight.

Never take the watch into an environment that exceeds the watch's water-ingress limitations. If the watch is capable of some degree of water resistance, remember to screw in the crown (or ensure it is pushed in) before immersing in or exposing it to water. Remember, your watch is a precisely engineered micromachine. It requires and deserves thoughtfulness and care.

STARTING A WATCH COLLECTION

"Time crumbles things; everything grows old under the power of Time and is forgotten through the lapse of Time."

—Aristotle

Like all collecting ventures, starting a watch collection begins with a budget and a desire to collect. Collecting begins with a single watch, though it is possible to imagine a virtual collection: those watches worthy of purchase if one had the opportunity and the means.

The hobby of watch collecting is an extremely personal undertaking. Collections usually originate around a theme, a price point, a brand, or—for some—a haphazard and random assortment without any intended purpose at all, other than the ephemeral pleasure of ownership.

The majority of watch collectors collect based around a theme or brand (or both). Themes may include the following:

- Watch material (e.g., only precious metals such as platinum, rose gold, yellow gold, or white gold, or only steel, ceramic, or titanium)
- Band material (e.g., only watches with leather straps)
- Functions (e.g., only watches with chronographs, perpetual calendars, minute repeaters)
- Specialized functions (e.g., some collectors collect only flyback or split-second chronographs)
- Dial color (e.g., only white or black dials)
- Dial markers (e.g., only Arabic or Roman numerals or baton markers)
- Purpose (e.g., only dive watches or aviator's watches)
- Embellishments (e.g., only watches with diamonds or precious gemstones)
- Vintage watches, with some collectors focusing on a particular decade or era.
- Origin (e.g., only watches from Japan, Germany, or Switzerland)

Some collectors purchase watches only from a particular brand and, in some cases, only limited-edition or limited-production models from these brands. There are those who even restrict their collections further by purchasing only certain limited-edition precious-metal watches from a singular brand (e.g., only limited-edition platinum A. Lange & Söhne timepieces).

I personally believe that the most important objective to keep in mind—unless you are purchasing solely for investment purposes (a pursuit requiring specialized expertise)—is to love the watch(es) you buy.

Some collections start out with no clear intent other than personal enjoyment, but this evolves as the owner's preferences change. Sometimes, the collector starts out with the intention of collecting only one type of watch, but with time one's desires and tastes change, and the collection may change to reflect this fine-tuning.

Some collectors routinely sell part of their collection as they update or refine their collecting approach.

There is no rule to collecting, but if investment is a primary or even essential reason for collecting, it is wise to consider the preferences of the majority of the watch-collecting community before making a purchase. That way, when it comes time to resell, you are in the best position to benefit.

Although the watch is the ultimate reason for collecting, its accoutrements may be just as valuable to collectors and, in some cases, may mean a realized profit differential of tens of thousands of dollars due to nostalgic significance or an ability to establish provenance. These items include the original box, the certificate of authenticity (if available—it isn't often), hangtags, the instruction manual, and sometimes even the original store receipt.

Most modern collectors prefer watches unrestored, with original (and often faded or oxidized) luminous material, and the watchcase in as-is condition, no matter the state of wear and tear. A watch in its original condition helps validate its bona fides and prevents fraudulent claims. A watch incorporating only original or factory-certified parts is also often a preference.

Purchasing a watch from a reputable auction house or dealer goes a long way to ensuring its authenticity but does not guarantee it. There have been several high-profile examples of auctioneers or dealers unwittingly selling watches that were less than claimed.

At 39 mm, the platinum A. Lange & Söhne Datograph Flyback Ref. 403.435 (ca. 1999, last known MSRP: $121,800) is both elegant and sporty.

If you're a collector of platinum watches, you can put checkboxes next to "platinum case" and "platinum bracelet" (a rarity). If you're a collector of "manual-wind movements," check! Also put checkmarks next to "flyback chronograph," "big date," and "German manufacture."

A. LANGE & SÖHNE
DATOGRAPH
FLYBACK
GLASHÜTTE
GERMANY
BASE 1000 METERS

OTHER FACTORS TO CONSIDER WHEN PURCHASING

If I had a dollar for every time someone asked me to recommend a watch, I would have $216, or thereabouts. The answer is surprisingly tricky.

What are your preferences and requirements? That is usually my first response. Are you looking for a watch that you can wear to work, the opera, or the shower; playing tennis; skiing in Aspen; dive training in Cozumel? All of the above? There are quite a few examples of watches that you can wear to all venues, but there is always a trade-off. Watches with higher water resistance are usually thicker, because they need to withstand extreme pressure and temperature. They require thick (or thicker) gaskets to seal the crown, pushers, case back, and crystal. See the subsection "Water Resistance" below for more on the subject.

You have decided you need to have a single watch that can do it all—or most of what you need; that way, you can also budget for only a single timepiece and put all your resources into one horological possession. If you want your watch to do it all and have a degree of longevity before an issue arises, I believe you automatically disqualify rubber, fabric, and leather straps (unless you plan to store the original and use a backup strap for normal wear). I do not know a single watch brand that can build a non-metallic bracelet capable of withstanding five years of mixed usage—two or three years at the absolute limit, but not five years. Rubber wears out, then cracks or tears (or both). Leather straps will give out after about a year of regular wear—even original straps from the manufacturer. Leather cracks and discolors, especially for those who perspire a lot or work out with their leather-banded timepieces. "Waterproofed" leather (leather treated to be hydrophobic[223]) is a better alternative to its untreated brethren, though the supple feel and texture of the leather usually suffers considerably.

After purchase, many collectors immediately switch out generic leather, fabric, or rubber straps to preserve the original for potential resale later.

Soft-metal bracelets are also mostly off the table as daily options, unless you are the kind of person who can be particularly careful with your possessions. I assure you, doorknobs, faucets, and corners of furniture do not care how careful you think you are. Soft metals such as 18K rose, white, or yellow gold will dent more easily than stainless steel or other hardened metals, and gold is really heavy (more than twice as dense as steel), so large gold watches are sometimes too weighty for smaller wrists.

Titanium generally serves double duty, being both tough and light. The metal may appear slightly "greenish brown" in color, so aesthetically, if the aerospace color scheme were less to your liking, stainless steel would be a more thoughtful choice. Steel, however, is a great deal heavier than the equivalent watch in titanium. Titanium is 45 percent lighter than steel.

White gold, platinum, and palladium are excellent choices for the person who wants to own a precious metal watch but also maintain a low profile with a watchcase that could easily be mistaken for stainless steel. Platinum is even heavier than gold, being two and a half to three times denser than steel. Platinum watches, even on leather straps, are incredibly hefty, and only those individuals without back problems should venture down that path.

Is legibility important? Large numerals or stick markers and a clean contrasting dial will be good choices for those needing instant accessibility to the current time. Do you need to know the time at night? At all hours of the night? If knowing the time in the dark (say, at 4 a.m.) is important to you, applied tritium tubes are probably close to the only choice. Few watches with applied luminous paint can provide sufficient light many hours after bedtime, while tritium tubes always shine at the same brightness (notwithstanding their half-life—see earlier discussion on this topic). However, tritium is generally duller than fully lit Super-LumiNova. For repeated practical use in dark environments, at the very least it is important to

223. Hydrophobic materials repel water.

have good luminous material on the hour, minute, and second hands. During the day, large hands and markers make for much easier reading too.

Is it essential to know the date? With the surfeit of calendars in your life, is such a feature truly necessary? If you need a date on your timepiece, do you also need to know the day? How is your vision? Poor? Consider a magnified date.

If you need even more than this—say, a watch with a calendar that requires servicing only once per century, is also mechanical, and offers sufficient water resistance and shock protection—you are in rarefied territory. Such perpetual-calendar dive watches are usually available only above $20,000.

Do you do a lot of traveling? Does this travel involve several time zone changes? Consider a watch with multiple time zones.

Going diving only occasionally? Will you use your watch as a backup? If not, it is probably unnecessary to have a unidirectional rotating bezel.

For me, the type of material used for the front glass of the watch is nonnegotiable. If made of mineral glass (even *hardened* mineral glass), your watch crystal will inevitably scratch, chip, or crack. Sapphire crystal is an absolute must, especially if servicing the timepiece within the first few years is out of consideration.

If the sapphire crystal boasts an exterior-facing antireflective treatment, take care to avoid using abrasives or harsh cleaning materials on the outer surface of the glass—forever. In my opinion, unless the reflective coating is permanent (or necessary because of the particular environment in which the watch will find use), it is probably better to choose a crystal without outer antireflective treatment. Otherwise, regular refurbishment of the coating is inevitable.

A crown protector is useful to prevent the crown from knocking against objects and dislodging or creating a bent stem. Crown protectors are watchcase protuberances above and below the crown, designed to divert knocks and shocks. Screw-in crowns (with or without crown protectors) are best, since they also offer enhanced water resistance, but they do tend to be annoying to screw and unscrew constantly, especially if the watch is manual winding.

Do you want your watch to work every morning, whether you took it off at the start of a weekend or not? You are looking for an automatic (self-winding) timepiece. These feature a rotor that rotates with your motion and "charges" the mainspring.

Left off the wrist, some automatic mechanicals can maintain their power reserve for up to a week or more. More enamored with the look of the movement? A hand-wound timepiece is perhaps just your speed, and you will probably want to purchase a timepiece with an exhibition case back too, so you can see the unobstructed majesty of the movement; mechanical automatics have rotors that obscure the beautiful view of the movement.

However, the one huge downside is that you will have to wind a mechanical movement routinely depending on the amount of inbuilt power reserve. A thirty-six-hour power reserve will necessitate winding once every day and a half. A ninety-six-hour power reserve will give you four-day winding intervals. If you want to sit back, relax, and not worry about when or where you take off your watch and how often you put it back on, then you are looking for a quartz watch. Wristwatch cognoscenti may be mocking of timepieces featuring quartz movements, but some single-malt-only Scotch drinkers mock their blended brethren and yet a large and dedicated following in the blended Scotch world continues to thrive. See the section "The Watch and Its Movement" for more details on differences between these movements.

WATER RESISTANCE

A watch rated at 30 meters of water resistance may withstand a static pressure test at 3 atmospheres (equivalent to 30 meters) for short stints, but it would not tolerate continued usage for swimming despite the swimming pool's depth being far less than 30 meters at its deepest point. In a diving or active swimming environment, additional factors come into play for which lower-rated water-resistant watches are unsuitable. The quality of the water (fresh water, seawater, water temperature), the rate of change of depth, and other dynamic factors will affect the actual water resistance.

International standards provide a degree of comfort regarding water resistance. ISO 2281, introduced in 1990, and its successor ISO 22810:2010 cover watches intended for normal daily wear.

ISO 6425 is the relevant diving-watch standard. To meet this standard's stricter criteria, the watch must withstand depths of 100 meters or more.[224] Depending on one's intentions for wearing the watch, it may be essential for the timepiece's usage to match or exceed its certified rating. If you intend to go diving, it is best to choose a watch with greater than 200 meters of water resistance or ISO 6425 certification.

If you are intent on buying a dress watch and its water resistance is less than 30 meters / 100 feet, consider it wise to remove it before exercise or prolonged water exposure. In my opinion, 30 meters may as well read "splash resistant," and indeed many brands use this term for water resistance of 30 meters or less.

Please refer to the section "Watch Styles: Dive/Nautical" for a more comprehensive discussion on water resistance.

Rolex Submariner 114060 (last known MSRP: $7,900, ca. 2012). The one to rule them all. This 300-meter water-resistant timepiece has a reputed water resistance of 100 meters even **with the crown open**.

It is legible in all conditions. The dial exhibits a simple but effective layout sans date (a date-based version is also available), and the case lines are clean and crisp.

The 904L corrosion-resistant steel and classic design make the Submariner the optimal choice for the luxury watch buyer looking for a new timepiece that serves multiple roles. Resale values tend to be strong over time too. **There is a reason that Rolex is the king of watches, and this is Exhibit A.**

224. The watch must conform to at least 125 percent of the rated diving depth. Unlike ISO 2281, which allows for sample testing, ISO 6425 requires every engraved example to be tested. The watches must also be able to withstand a depth of approximately 30 cm for fifty hours at a range of temperatures (18°C–25°C) and continue functioning normally. A condensation test heats the watch on a plate to between 40°C and 45°C. A droplet of water at a temperature of 18°C–25°C drips onto the glass. After a minute, if no condensation appears on the inner surface of the glass, the watchcase passes. Additional water pressure tests and immersion (back and forth) in water of warm and cold temperatures satisfy the remaining test requirements. Mechanical dive watches also require a method for time preselection (e.g., a unidirectional rotating bezel). The time, bezel, and running seconds should also be visible in the dark at a distance of 25 cm (about 10 inches). Magnetic and shock resistance appear in the list of requirements too, together with a resistance to salt water. Some watches also feature a helium escape valve, which is necessary for mixed-gas diving. These watches require additional testing requirements.

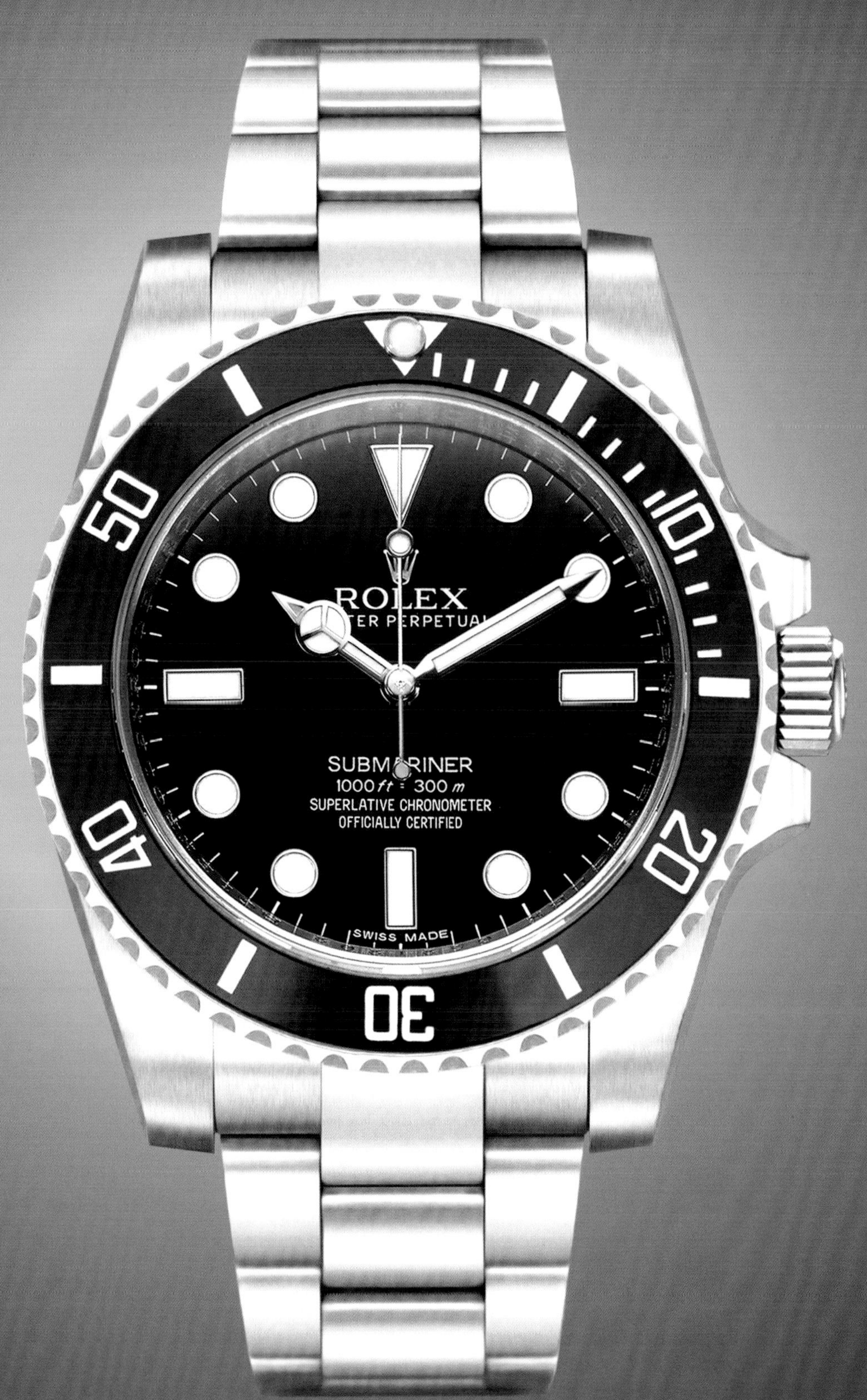
ROLEX
1000 ft = 300 m
SUPERLATIVE CHRONOMETER
OFFICIALLY CERTIFIED
SWISS MADE
10
20
30
40
50

Harry Winston Opus 12 (Ref. 500/MMEB46WL.K, $260,000 in 2012, limited to 120 pieces). An excellent choice for an offbeat dress watch. Case in polished and brushed 18K white gold. This novel tour de force consists of 607 parts, eighty jewels, and **twenty-seven hands**. If traditional time telling is too uninspiring for you, the Opus 12 by Emmanuel Bouchet (in collaboration with Harry Winston) is one solution.

Every five minutes, the retrograde hand—the long blue hand at the dial center—transits across the graduated five-minute hemisphere. Then, one of the twenty-four blue/silver hands flips over to display the current five-minute interval (longer blue hand on the periphery). Every hour, all twelve hour hands perform an animated "dance sequence." The time in the illustration is 08:21:45. At 08:25, the minute hand at five o'clock (by the letter "T" on the flange) will flip to blue, and the minute hand at four o'clock (by the letter "O" on the flange) will flip to silver. The two other hands are the running seconds and power reserve indicator.

HARRY WINSTON
0 1 2 3 4 5
SWISS MADE

THE INNER WORKINGS OF THE WATCH INDUSTRY

"Let every man be master of his time."

–William Shakespeare, *Macbeth*

WATCH MANUFACTURING

An intriguing aspect of horology is the behind-the-scenes workings of an interconnected industry where the supply chain is both extensive and convoluted. The watch industry started as a cottage industry and remained relatively unaffected until the Industrial Revolution; even then, due to antiquated production techniques and bureaucratic guilds, many parts of the industry took decades to industrialize effectively.

Many other books provide well-researched coverage on the early years of watchmaking, especially Swiss, British, German, American, and Japanese watchmaking. Brand-specific tomes delve into the intricacies of brand formation and the surrounding watchmaking environment. Thus, I provide a very brief overview here and encourage further exploration regarding watch production histories of specific countries, geographical regions, or brands.

A BRITISH SPECIALTY MASTERED BY THE SWISS

Watch manufacturing was largely a British specialty throughout the seventeenth and eighteenth centuries, but production methods supported only small-batch, high-quality (at least for the period) timepieces. Contemporaneous Swiss watchmaking demonstrated a similarly limited output. The existing guild system, content with its dictatorial structure, stood in deliberate opposition to attempts at mechanization and industrialization. However, toward the middle of the eighteenth century, especially due to the concerted efforts of the artisan Daniel JeanRichard applying division-of-labor principles,[225] many local farmers learned and mastered specialized production techniques, especially useful expertise during Switzerland's cold winter months. By the end of the eighteenth century, the Swiss were exporting tens of thousands of watches, and JeanRichard's subcontracting system of specialization had spread throughout Switzerland and birthed hundreds of attendant specialist professions.

The British attempted timepiece industrialization, but the effort proved challenging. By 1861, however, the Americans had mostly worked it out. The then-ten-year-old Massachusetts-based Waltham Watch Company was running a successfully industrialized factory utilizing interchangeable parts and providing improved accuracy, especially necessary for a rapidly expanding railroad industry that required precision for train and passenger safety.

225. Adam Smith, *An Inquiry into the Nature and Causes of the Wealth of Nations* (1776). In his seminal work, Adam Smith, the pioneering Scottish economist, philosopher, and author, describes the rudiments of industrialization via the process of pin making: "*One man draws out the wire; another straights it; a third cuts it; a fourth points it; a fifth grinds it at the top for receiving the head; to make the head requires two or three distinct operations; to put it on is a peculiar business; to whiten the pins is another; it is even a trade by itself to put them into the paper; and the important business of making a pin is, in this manner, divided into about eighteen distinct operations, which, in some manufactories, are all performed by distinct hands, though in others the same man will sometimes perform two or three of them. I have seen a small manufactory of this kind, where ten men only were employed, and where some of them consequently performed two or three distinct operations. But though they were very poor, and therefore but indifferently accommodated with the necessary machinery, they could, when they exerted themselves, make among them about twelve pounds of pins in a day. There are in a pound upwards of four thousand pins of a middling size. Those ten persons, therefore, could make among them upwards of forty-eight thousand pins in a day. Each person, therefore, making a tenth part of forty-eight thousand pins, might be considered as making four thousand eight hundred pins in a day. But if they had all wrought separately and independently, and without any of them having been educated to this peculiar business, they certainly could not each of them have made twenty, perhaps not one pin in a day.*"

In 1876, the first international watch precision competition ran during the International Centennial Exposition in Philadelphia (the first official world's fair held in the United States). At the exhibition, Waltham displayed the first automatic screw-making machinery, while simultaneously taking home the gold medal for watchmaking precision. The traditional watch industry has come a long way since then, with the Swiss currently occupying the dominant industry position.

INDUSTRY SIZE

SWISS WATCHES

The Federation of the Swiss Watch Industry (FHS) publishes monthly statistics approximately three to four weeks after the close of the prior month. As the watch industry's trade association, the FHS provides public data going back to the year 2000. World luxury Swiss watch exports[226] totaled CHF (Swiss franc) 9.277 billion in 2000.[227] By 2010, that value had grown to CHF 15.154 billion. For 2019, the figure was CHF 20.503 billion. This record of financial growth looks deceptively promising, at least prior to 2020 (before the effects of COVID-19). Swiss watch exports in 2020 were especially depressed, given the enormous logistical and economic challenges brought about by the pandemic. Full-year exports for 2020 both for wristwatches and movements were a gloomy CHF 16.984 billion (CHF 16.124 billion in completed watches alone), setting the industry back almost a full decade in terms of revenue appreciation.

Even before the onset of COVID-19, 2019's results hid an underlying structural uncertainty that still pervades the industry. More on that later.

The biggest market for luxury wristwatches is Asia. From a *mere* CHF 3.9 billion in 2000, in 2019 Asia accounted for around CHF 11.5 billion[228] (of Swiss watch exports). Newly minted millionaires and billionaires from Asia have transferred their vast wealth into wristwatches, both as a sign of status and as a potential investment vehicle.

In Asia, recent and vintage Patek Philippe, Rolex, Audemars Piguet, and Richard Mille timepieces command disproportionate shares of the luxury watch industry, and as a result, these brands have put much of their resources into satisfying the needs and demands of this heated market.

More than half of the entire Swiss luxury watch industry's output (by value) finds a home in Asia, and that is merely by taking the export data at face value. Transshipping from other territories probably adds another several billion dollars to that already outsized amount. (Transshipment is the shipment of goods to an intermediate destination and then onward to another destination—in the watch industry, this usually takes place without the knowledge or approval of the watch manufacturer or brand holder.)

I discuss the parallel importation of watches in more detail later.

As of this writing, 2020 is the most recent year of complete annual historical data for the Swiss watch industry. However, the deleterious economic impact of COVID-19 makes for a difficult comparative examination of 2020's data relative to prior years. Adjusted financial results for 2019 show watch exports at CHF 20.502 billion (2.8 percent up over 2018). The slight uptick in sales appears encouraging at first glance, but a much more worrying picture emerges when performing analysis of volumes and price points.

Sales volume was down in 2019. Way down. Certainly in comparison to two decades prior. Units exported in 2000 were 29.656 million. In 2019, there were recorded exports of only 20.646 million units—a decline of 31 percent—largely attributable to the attrition of the sub-200 (CHF) price point. Units exported in this category declined by a rather striking 49 percent from 2000 to 2019. Much of the lost value from these lower price points is made up for in the price point over CHF 3,000, where exports increased from 488,000 units (2000) to 1.668 million units (2019). The concomitant value increase was a staggering CHF 11 billion. In this price category, the average watch exported from

226. Exports described are for completed watches only and do not include movement exports (which would make the values marginally higher).

227. CHF is the ISO 4217 currency code for the Swiss franc. The value of the Swiss franc to the US dollar has fluctuated within about 10 percent of parity in the previous five years, so for ease of currency conversion, consider one dollar as equal to one Swiss franc.

228. "According to the Federation of the Swiss Watch Industry, the most recent figures show exports of Swiss watches to mainland China totaled 2.1 billion Swiss francs ($2.39 billion) from January to November 2020, an increase of 17.1 percent from the same period in 2019. (The federation tallies exports to Hong Kong and to mainland China separately.) Every other market in the industry's top 20—including Hong Kong and the United States, its longtime leaders—recorded a double-digit decline during the same period."
Robin Swithinbank, "China Takes the Lead in Demand for Swiss Watches," *New York Times*, January 14, 2021, https://www.nytimes.com/2021/01/14/fashion/watches-china-swiss-export-leader.html.

Switzerland carried a value of just over CHF 8,501 in 2019 (just shy of 2015's record average of CHF 8,523). For comparison, the average export value in the "above CHF 3,000" segment in 2000 was CHF 6,461. Thus, in two decades, prices in the "CHF 3,000 and above" segment increased by approximately 31 percent in average export value; volumes increased about 242 percent for the same period. (See chart: Swiss Watch Exports 2000–2020).

What does it all mean?

It means that market forces are shifting consumption patterns in the luxury watch industry, and that these changes are significant and rapid—made ever more urgent by COVID-19. Data for 2020 shows significant declines against 2019 (down 21 percent for overall finished watch export value), but isolating the pandemic effect will prove challenging. The sub-200 CHF category's performance continues eroding due to stiff competition from readily available Asian equivalents, changes in watch-wearing habits, the rapidly rising dominance of smartwatches and smart devices, and changes in the cultural zeitgeist.

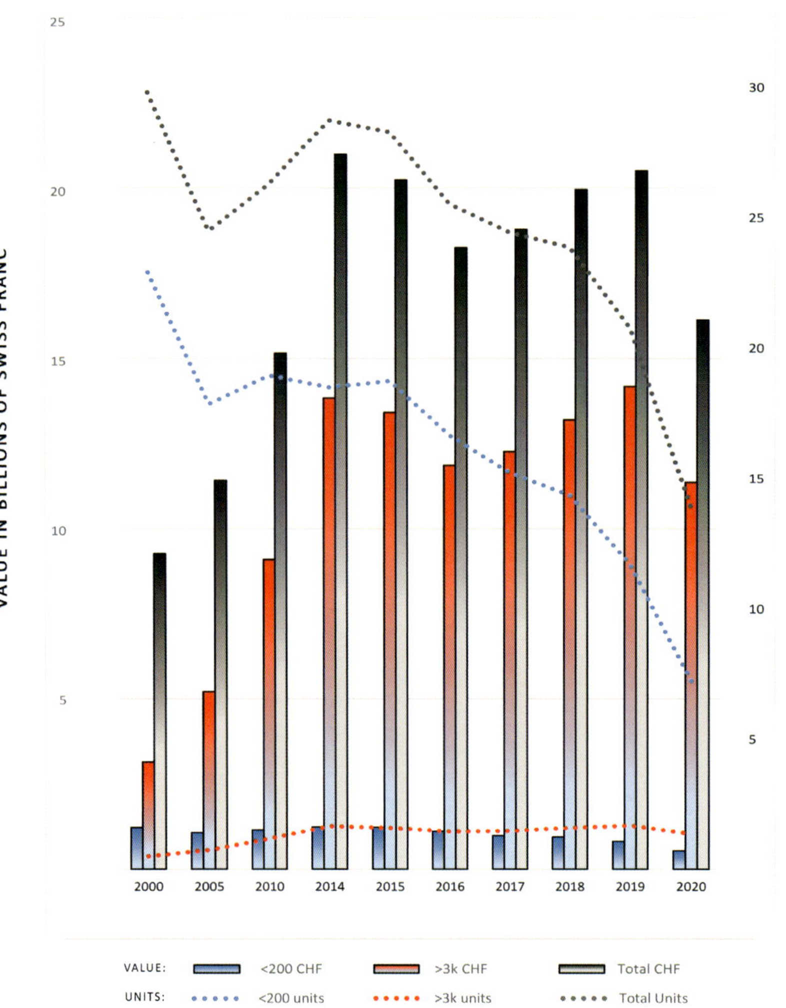

Data source: Federation of the Swiss Watch Industry FH

The only apparent bright spot for the Swiss industry in 2019 was the overall export value increase in the "CHF 3,000 and above" price point. Brands producing timepieces with an export value less than CHF 3,000 (equivalent to about CHF 10,000 at retail) are at risk. The value statistics presented here are at export pricing, which is roughly equivalent to distribution value (before tariffs and other associated costs), and should not be confused with overall worldwide *retail revenue* for Swiss watches, which is several multiples larger.

Year	Value in millions	Price point in CHF			
		<200	<200–500	<500–3000	>3000
2000	9,277	1,231	1,036	3,856	3,153
2005	11,418	1,072	841	4,286	5,219
2010	15,154	1,160	1,043	3,858	9,094
2014	20,988	1,249	1,559	4,344	13,837
2015	20,238	1,237	1,425	4,161	13,415
2016	18,257	1,112	1,289	3,998	11,858
2017	18,789	983	1,343	4,204	12,260
2018	19,948	946	1,299	4,517	13,187
2019	20,502	818	1,227	4,278	14,179
2020[229]	16,124	541	817	3,407	11,359

Table: Value (in Millions of CHF) of Swiss Watches Exported by Year and Price Point[230]

Year	Volume in thousands	Price point in CHF			
		<200	200–500	<500–3000	>3000
2000	29,656	22,795	3,144	3,229	488
2005	24,364	17,769	2,574	3,291	731
2010	26,148	18,853	3,215	2,920	1,161
2014	28,586	18,400	4,997	3,546	1,643
2015	28,138	18,644	4,506	3,414	1,574
2016	25,396	16,551	4,115	3,292	1,438
2017	24,305	15,135	4,301	3,408	1,461
2018	23,741	14,286	4,189	3,689	1,577
2019	20,634	11,621	3,964	3,382	1,666
2020	13,774	7,203	2,594	2,641	1,340

Table: Volume (in Thousands) of Swiss Watches Exported by Year and Price Point[231]

229. Cumulative values for 2020 were obtained via monthly aggregation (this applies to all 2020 FH data—see also below) and may vary with FH's final tallies due to rounding and any amendments (if applicable).
230. Source: Federation of the Swiss Watch Industry FH.
231. Ibid.

Year	Average Price in CHF	Price point in CHF			
		<200	<200–500	<500–3000	>3000
2000	313	54	329	1,194	6,461
2005	469	60	327	1,302	7,139
2010	580	62	324	1,321	7,833
2014	734	68	312	1,225	8,421
2015	719	66	316	1,219	8,523
2016	719	67	313	1,215	8,246
2017	773	65	312	1,233	8,391
2018	840	66	310	1,224	8,362
2019	993	70	309	1,265	8,490
2020	1,171	75	315	1,290	8,476

Table: Average Price (in CHF) of Swiss Watches Exported by Year and Wholesale Price Point[232]

CHINA AND HONG KONG

The Swiss watch industry's weakening demand isn't isolated. China produced 644.3 million watches for export in 2019[233]—6.4 percent fewer than in 2017. Moreover, the average price per unit is declining: from $4 in 2017 to $3 in 2019. Hong Kong exported 13.2 percent fewer watches in 2019 than it did in 2017, when it produced 227.9 million units for worldwide consumption.

SMARTWATCHES

IDC's Worldwide Quarterly Wearable Device Tracker[234] forecast smartwatch sales at 69.3 million units for 2019 (22.7 percent of all wearable devices) and increasing to 109.2 million units in 2023 (22.3 percent of all wearable devices). Smartwatches offer a compelling alternative to a traditional analog or digital watch. A powerful example is the growing awareness of personal health, which is stimulating demand for smartwatches that measure biological processes. These devices now also offer cell phone connectivity, GPS, distance tracking, Bluetooth connectivity, and much more.

As consumers optimize wrist real estate in line with the flourishing smartwatch trend, traditional watch sales will be dramatically affected, especially in the lower-priced categories.

232. Ibid.
233. Ibid.
234. December 16, 2019.

COVID-19

While several vaccines became available toward the tail end of 2020, it still took time for economic activity to return to a semblance of normal.

For 2020, total watches exported from Switzerland decreased to 13.8 million units from 20.6 million units in 2019. Export value declined to CHF 16.1 billion (from CHF 20.5 billion in 2019). Swiss wristwatch exports were 33 percent lower year over year by unit volume and down 21 percent by value over the same period.

All but one of the top twenty export markets declined, excepting the largest (China), which grew an astonishing 20 percent over 2019.[235] Year-over-year watch export values declined by over 17 percent in the United States (the second-largest market), 37 percent in Hong Kong, 26 percent in Japan, 24 percent in the United Kingdom, 26 percent in Singapore, 21 percent in Germany, 18 percent in United Arab Emirates, 9 percent in France, and 33 percent in Italy (rounding out the top ten). Hardest hit of the top ten markets was Hong Kong (lower by 37 percent), with a decline in export value from CHF 2.691 billion in 2019 to CHF 1.697 billion in 2020.

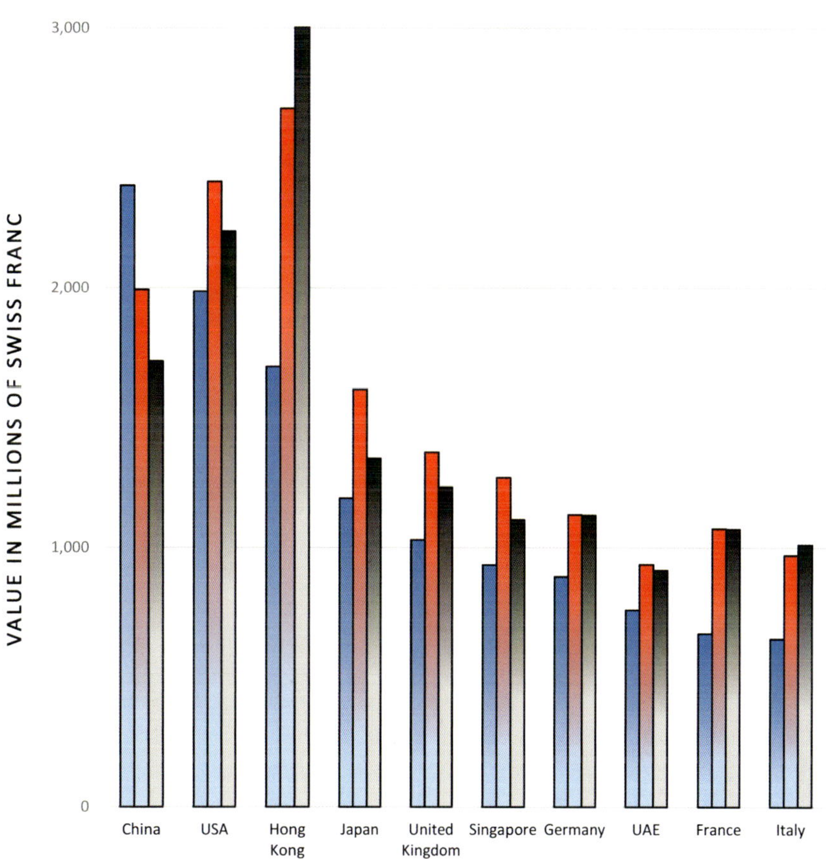

Data source: Federation of the Swiss Watch Industry FH

235. See chart: Top 10 Swiss Watch Export Markets.

The A. Lange & Söhne L951.1—the 95 designation is to indicate that development began in 1995—looks like a miniaturized metal city from above. One could stare at it for hours. The immaculately chamfered parts, blued screws, and awesome three-dimensionality make it a personal favorite.

A. LANGE & SÖHNE
GLASHÜTTE I SA.
REGULIERT
IN FÜNF (5) LAGEN
VIERZIG
(40) RUBINE

Apart from the pandemic effects, the political protests and Chinese government intervention in the "Special Administrative Region" engendered additional negative repercussions for the industry.

BUILD VS. BUY

Does a watch brand make its own movement, or does it outsource the movement to a specialized manufacturer? This decision became even more pressing when COMCO (the Swiss Competition Authority) authorized Swatch Group's ETA to reduce movement supply to the watch industry over the course of nearly a decade. By the third quarter of 2020, ETA's COMCO obligations to other suppliers had finally expired. ETA, the largest Swiss mechanical movement supplier, traditionally supplied off-the-shelf movements to numerous brands, but as of 2021, suppliers such as Sellita (with "ETA-equivalent" movements) provide the majority of nongroup lower-tier watchmakers with their movement requirements.

An in-house-developed movement definitely provides a brand with a degree of bona fides. Yet, it is not always cost effective or practical to develop a movement in-house. On the basis of independent research, even a relatively simple movement would cost at least $6 million to develop and mass-produce.

MOVEMENT MANUFACTURERS

Not all movements are created equal. Just as car engines vary in design, build quality, efficiency, power output, and reliability, so do watch movements. Many luxury watch movements are still hand-finished, but with the prevalence of twenty-first-century manufacturing techniques, including 3-D printing and CAD/CAM manufacturing, most modern movements, even those at entry-level price points, provide excellent workmanship and outstanding value.

Only a handful of well-known watch brands nestle at the pinnacle of movement artistry. These brands offer a high degree of movement complexity, specialized componentry in terms of their physical and mechanical properties, and sophisticated hand finishing on various movement parts. The dedicated connoisseur may know most of the names on this list. I have not separated in-house movement makers from commercially available movement makers. This is also not an exhaustive list, nor does it include exclusively quartz movement makers. Last, the decision to include a movement manufacturer in a particular segment (e.g., superlative vs. outstanding) is subjective and certainly debatable.

Superlative movement makers (in alphabetical order): A. Lange & Söhne, Credor, De Bethune, F. P. Journe, Ferdinand Berthoud, Grand Seiko, Greubel Forsey, Grönefeld, H. Moser & Cie, Laurent Ferrier, MB&F, Moritz Grossman, Patek Philippe, Richard Mille, Urban Jürgensen & Sonner, Voutilainen

Outstanding movement makers: Antoine Preziuso, Armin Strom, Audemars Piguet, Breguet, Bvlgari, Christiaan van der Klaauw, Christophe Claret, Czapek Genève, Girard-Perregaux, Glashütte Original, Jaeger-LeCoultre, Piaget, Roger Dubuis, Ulysse Nardin, Urwerk, Vacheron Constantin, Vianney Halter

Excellent movement makers: Akrivia, Blancpain, Bovet, Breitling, Carl F. Bucherer, Cartier, Chopard, Corum, Eterna, Fabergé, Franck Muller, Habring[2], Hermès, Hublot, HYT, IWC Schaffhausen, Jaquet Droz, Konstantin Chaykin, Louis Moinet, Maurice Lacroix, Montblanc, Nomos, Omega, Panerai, Parmigiani Fleurier (Vaucher), Rolex, TAG Heuer, Tudor, Van Cleef & Arpels, Vaucher Manufacture, Vulcain, Zenith

Other movement makers: Concepto, ETA, La Joux-Perret, Sellita, SOPROD, Miyota, Seiko

On a personal note, my favorite movement is the hand-wound, 32 mm, forty-jeweled, 405-component A. Lange & Söhne L951.1 (ca. 1999). Lange manufactures movements that are certainly more complex—consider the L952.2, with 729 parts, or the L101.1, with 631 parts. However, the first time I observed the L951.1 in the Platinum Datograph Flyback Chronograph (Ref 403.035), it became an instant favorite. There are a few Richard Mille movements (some Greubel Forseys and many others too) that offer similar three-dimensional magnificence. However, for me, the refined complexity, superlative layout, and extraordinary finishing on the Lange movement make for an easy choice as my number one selection.

DISTRIBUTION

TRADITIONAL DISTRIBUTION MODEL

The traditional watch distribution model involves manufacturer, country agent/distributor, wholesaler (a relatively rare additional level), retailer, and consumer. In a direct-to-consumer model (DTC), nascent brands have eliminated the distribution and wholesaling strata. Instead, they sell the product directly to the end consumer.

An example may help elucidate. Typical luxury watch manufacturers generate approximately 67 percent gross margin.[236] Using this statistic as a benchmark, if the brand manufacturer produces a watch for $100, it would sell it to the distributor for $300. The gross margin is the difference between these two amounts as a percentage of the distribution amount. Large distributors or country agents typically buy a luxury watch for between 70 and 75 percent off the manufacturer's suggested retail price (MSRP). Using some basic math and an estimated benchmark of 75 percent off MSRP for the distributor, we arrive at a retail price of $1,200 ($300/0.25). The "profit" of $900 apportions among the distributor, the wholesaler, and the retailer. To keep things simple, for this example we will not utilize a wholesaler.

The distributor is responsible for a large percentage of the marketing costs in its local market and typically shares the high costs of advertising and promotion with wholesalers and retailers. The retailer will traditionally garner 35–50 percent of the final selling price. If we use 40 percent for our example, then $480 of the retail value would go to the retailer. The remainder is profit for the distributor. In this example, the remainder is $420 ($1,200 – $480 – $300). Thus, a watch manufactured for a mere $100 necessitates a retail price of $1,200 in order to satisfy the "standard" margin demands of distributor and retailer.

Startup brands have realized that "cutting out the middleman"[237] removes two levels of markup, and the $100 watch sells directly to the consumer for $300. The $900 distribution profit/cost evaporates and the customer benefits by paying a lower price. At least in theory.

The challenge for existing brands is navigating the DTC model while continuing to accommodate traditional distribution models.

PARALLEL IMPORTING/GRAY MARKET

The World Trade Organization's glossary of terms defines "parallel imports" thus:

> *When a product made legally (i.e., not pirated) abroad is imported without the permission of the intellectual property right-holder (e.g., the trademark or patent owner). Some countries allow this, others do not.*

Importantly, these goods are legal, albeit surfacing for sale without the *official*[238] imprimatur of the brand holder. For decades, the Swiss wristwatch market, in addition to many other markets for items with nonarbitraged asset values—think pharmaceuticals, cell phones, and electronics—has operated with a relatively healthy parallel market. Simply put, gray or parallel markets exist only in industries where global pricing is nonuniform (currency and tariff adjusted).

WORKING EXAMPLE:

A Rolex Daytona 116500 LN watch sells for 12,250 euros in Germany.[239] Assuming you could actually walk into a retailer and purchase one—you cannot: waiting lists, even when not completely filled up, are notoriously long, often extending beyond several years—you would pay $13,150 in the United States. In the United Kingdom, the MSRP is £10,500. The same timepiece would set you back ¥1,387,100 in Tokyo.

236. This margin percentage is an average taken across numerous well-known watch manufacturers. Some brands will have higher margins than others depending on their cost allocations and overhead structure. A good rule of thumb: watches are typically marked up at three times cost, yielding a 67 percent gross margin.

237. A phrase so overused in promotional material that it has become trite.

238. In downward economic business cycles, many brands "casually overlook" gray-market activities, enjoying the short-term increase and boost to group revenues that this parallel market facilitates. I am also aware of dubious brand-initiated invoicing practices involving selling items to the gray market (directly) at higher invoiced wholesale prices, with immediate concurrent offset discounts to conceal the true nature of the questionable transaction. While the complexity and extent of these practices are difficult to quantify, I have to laugh at the hypocrisy of such clandestine duplicity when the very same brands "declare war" on the gray market.

239. As of July 2020, including VAT, which may be refunded if the merchandise is purchased and exported by a customer whose residence is outside the European Union.

The cross-rate table below shows how many units of one currency convert into the other currencies—e.g., $1 (US) is equivalent to €0.874 (euro). The dashes show the currency compared with itself, which is obviously equivalent.

Cross-rate	EUR	USD	GBP	JPY
EUR	—	0.874	1.1065	0.0081
USD	1.1441	—	1.266	0.0093
GBP	0.9037	0.7899	—	0.0074
JPY	122.7612	107.28	135.8165	—

Table: Cross-Rates of Four Major Currencies as of July 20, 2020 (Source: Bloomberg)

Using the retail values for the Rolex Daytona, the cross-rate table transforms into an arbitrage[240] analysis table. Say, for example, that you were on vacation in Tokyo. Should you purchase a Rolex Daytona there? From a purely currency-based arbitrage position, and assuming that any saving above zero is worthwhile, the simple answer is yes. The equivalent price in US dollars is $12,900 in Tokyo against a localized price of $13,150 in the US (a savings of $250). Admittedly not much of a saving, but if instead of Japan you made a purchase in Germany (or elsewhere in the EU), the VAT inclusive amount reduces by 19 percent,[241] rendering a final cost to the visitor of $11,352.15 (81 percent of €12,250 multiplied by the exchange rate), a savings of $1,797.85. While Rolex tries its utmost to create global pricing parity (equivalence), were it not for routine local pricing adjustments, these values would quickly misalign.

	EUR	USD	GBP	JPY
Price (local currency)	12,250	13,150	10,500	1,387,100
EUR	**12,250**	11,493	11,618	11,236
USD	14,015	**13,150**	13,293	12,900
GBP	11,070	10,387	**10,500**	10,265
JPY	1,503,825	1,410,732	1,426,073	**1,387,100**

Table: Watch Price Arbitrage Values

240. Arbitrage is a process wherein an investor can generate a profit from simultaneously buying and selling a commodity (or other asset) in two different markets. If, after accounting for exchange rate and duty disparities, the pricing differential of the asset in both markets is sufficient to generate acceptable returns, a market arbitrage opportunity exists and is open to exploitation. In efficient markets, these opportunities should not exist.

241. Current VAT rate in Germany as of autumn 2020.

A Rolex timepiece is probably not the optimum example because the brand maintains strict control of unauthorized cross-border imports, most notably into the US (one may import only a single timepiece—more than one unit clashes with existing trademark law). However, for purposes of understanding product-pricing arbitrage, it is instructive.

Overlooking the potential trademark issues,[242] suppose you are an affluent individual. Further, for purposes of this example let us say that the Japanese yen declines in value to 150 yen per US dollar, and that it does so in an accelerated fashion over a period of mere days or weeks. Rolex would have insufficient time to adjust local pricing in Japan. You now purchase the watch in Japan for 1,387,100/150 = $9,247. This is a substantial saving over the local price of $13,150. Of course, there are import duties in the United States to consider, but all things being equal, the price in Tokyo is far less expensive than in New York. Market forces would eventually prompt Rolex to increase the price in Japan to inhibit these and similar arbitrage opportunities.

For some brands, however, currency arbitrage and transshipping are serious challenges. In the example given, I used retail pricing and only a single timepiece. Extending this example to distribution pricing (70 percent off MSRP or more) and to hundreds or even thousands of timepieces, one can see how a well-funded parallel importer can use currency fluctuations, import tariff discrepancies, and local market demand analysis to run a successful product arbitrage business.

This, in brief, is how the "unsanctioned" gray market operates. Watches are purchased from a country agent or distributor in country A (their intended market) and are transshipped to country B (usually unbeknown to the brand holder), where demand may be greater and local pricing higher. When one sees a watch online for 60 percent off and it is a brand-new, nondiscontinued item from a major brand, it was in all likelihood procured by the seller through one or more gray-market channels.

Gray-market watches do not usually ship with a manufacturer warranty; these documents are usually available only via authorized distributors. However, glossing over the warranty issue and issues of originality (there are many incredibly realistic counterfeit items available through less scrupulous merchants), savings may be substantial. Brands, especially those under the umbrella of the large groups, are (or at least appear) extremely vigilant about shutting down gray-market sources. Many brands have specific clauses written into their retail contracts that penalize transshipping and other gray-market activities. These penalties include loss of authorized dealership status, a requirement for the offending retailer to purchase back the watch at full retail price (not the actual discounted dealer price that was paid to procure the watch), and censure from future procurement.

THE BIG WATCH GROUPS

The luxury watch industry is predominantly Swiss-centric, but there are several luxury brands outside Switzerland, notably in Japan, Germany, and the UK. A ranking of watch groups is difficult. Does one do so by overall revenue or only by revenue attributable to watches? LVMH is by far the largest group by revenue (2020 revenue of €44.65 billion), but its brand portfolio extends well beyond watches, and watches account for *only* €3.356 billion (2020). Richemont reported revenues of €14.238 billion (March 2020), but their "Specialist Watchmakers" reported sales of *only* €2.859 billion. The Swatch Group is almost entirely made up of watchmakers, with 2019 revenues of CHF 8.243 billion.[243]

At the beginning of 2020, following earlier analyses in 2018 and 2019, Morgan Stanley and watch brand consultancy LuxeConsult analyzed 350 brands within the CHF 50 billion retail Swiss watch industry and produced their annual Swiss watch industry report. On the basis of their evaluation, they determined that seven brands were members of the exclusive "billion-dollar club"—that is, brands with more than $1 billion in annual revenue: Rolex followed by Omega, Cartier, Longines, Patek Philippe, Audemars Piguet, and Tissot. Richard Mille may join the club for 2021's report of the preceding year. Rolex commands an astonishing 23.4 percent of retail market share, with Omega next in line at 8.5 percent.

Over the next few pages, I discuss the major watchmakers and their prominent brands with a selection of timepieces that are emblematic.

242. Trademarks and the application of the "first sale doctrine" are beyond the purview of this book, but the case law is fascinating.
243. Swatch Group reported first-half 2020 sales of CHF 2.20 billion, a YOY decline of 43.4 percent.

SWATCH

The king of the watch groups in terms of watch-related revenues and portfolio of brands, the Swatch Group is the industry's titan. The group itself prefers to enumerate its brand holdings within four tiers:

Prestige and luxury range: Blancpain, Breguet, Glashütte Original, Harry Winston, Jaquet Droz, Léon Hatot, Omega
High range: Longines, Rado, Union Glashütte
Middle range: Balmain, Calvin Klein, Certina, Hamilton, Mido, Tissot
Basic range: Flik Flak, Swatch

Swatch also owns ETA[244] and Nivarox-FAR,[245] two companies critical for parts and movement supplies to the entire industry. Omega, Longines, and Tissot each sell more than $1 billion annually.

Breguet Type XXI Flyback Chronograph in 18K rose gold (Ref. 3810BR/92/9ZU, MSRP: $20,900). 42 by 15.2 mm. Mechanical automatic, forty-five-hour power reserve. Calibre 584Q (twenty-five jewels, 4 Hz). Water resistant to 100 meters.

Elapsed minutes display via a central hand that sits just below the chronograph second hand and measures up to sixty elapsed minutes on the flange. Day/night indicator at three o'clock. Running seconds at nine o'clock. Twelve-hour register at six o'clock. Trapezoidal date aperture at six o'clock. Rose gold deployant clasp. Coin edge case band. Bidirectional bezel. Diamond-shaped faceted markers. Solid case back with hallmarking and brand engraving.

Breguet may not be Swatch's largest brand, but it is the most historically significant.

244. ETA (ETA SA Manufacture Horlogère Suisse) designs and manufactures quartz, hand-wound, and automatic-winding mechanical ébauches and movements. The company is a wholly owned subsidiary of the Swatch Group.
245. Nivarox is a Swiss company formed by a merger in 1984 between Nivarox and Fabriques d'Assortiments Réunis. Nivarox is also the trade name of the metallic alloy used in producing hairsprings (attached to the balance wheel).

Breguet
SWISS MADE

Blancpain Le Brassus 42 mm men's watch in 18K rose gold, Ref: 4286P-3642A-55B (ca. 2009, MSRP: $67,700). Featuring split-second flyback chronograph and perpetual calendar (day, date, month, year, leap year) with moon phase.

Founded in 1735, Blancpain is the world's oldest registered watchmaker. Before Swatch Group's acquisition, it was under the stewardship of Jean Claude Biver, a watch industry luminary. He is famous for introducing the slogan "Blancpain has never made a quartz watch and never will."

Blancpain is especially renowned for its Fifty Fathoms dive model.

BLANCPAIN
SWISS MADE

Although Omega is likely Swatch Group's most profitable and recognized brand, the crown jewel in terms of watchmaking provenance is Breguet—a moniker almost synonymous with watchmaking. Abraham-Louis Breguet founded the eponymous brand in Paris in 1775. Since 1999, it has been a Swatch subsidiary. Breguet pioneered the tourbillon and produced the world's first self-winding watch as early as 1780. Even the invention of the wristwatch is commonly attributed to Breguet.

Omega Speedmaster Professional, Ref: 311.30.42.30.01.006 (MSRP: $6,350). It features a 42 mm stainless-steel case with domed anti-reflective sapphire crystal. Its almost identical cousin, the 311.30.42.30.01.005 (only the last digit in the SKU differs), features Hesalite (transparent thermoplastic) instead of sapphire for its glass. An iconic timepiece, the original "Moonwatch" Speedmaster is famous for its involvement in all six lunar missions and for its pivotal role in the safe return of the Apollo 13 crew.

Sapphire is much tougher than acrylic (also called Plexiglas, Lucite, and Perspex, among other trade names). Hesalite is Omega's trademark name for polymethyl methacrylate or PMMA. Sapphire will not show scratches because of its incredible hardness (second only to diamond). Hesalite, on the other hand, is four times softer than sapphire and will scratch relatively easily but benefits from a domed shape and a more transparent overall appearance, since sapphire may suffer from a cloudy ring effect in places.

Acrylic allows for more complex shapes than synthetic sapphire and can be injection molded, compression molded, and extruded relatively inexpensively. Sapphire requires lengthy machining with expensive tools to achieve similar results, and even when possible, machining the sapphire crystal may weaken its structural integrity in places. The astronauts took watches fitted with acrylic crystals to the moon—it was part of NASA's safety specification.

TACHYMÈTRE
OMEGA
Speedmaster
PROFESSIONAL

KERING

Kering reported revenues of €15.884 billion for the financial (and calendar) year-end 2019. Only 7 percent of its revenue is attributable to watches and jewelry.[246] Its watch brands include Boucheron, Gucci, Girard-Perregaux, Jean Richard, and Ulysse Nardin.

Girard-Perregaux's Ref. 8028 (ca. 2006). Limited to 150 pieces in 18K white gold, it includes a twelve-hour chronograph, tapered rubber strap, and 18K white gold deployant clasp. Girard-Perregaux's relationship with Italian supercar maker and Formula 1 champion Ferrari lasted from 1994 until 2004. This particular timepiece celebrates Ferrari's Grand-Am GT Championship in 2002 and 2003, commemorated with an insignia on the case back.

246. Kering.com, February 12, 2020.

GIRARD-PERREGAUX
360 GT
SWISS MADE

First released in 2012, Ulysse Nardin's avant-garde El Toro (Ref. 329-00), shown here in platinum and ceramic, features a quick-setting time zone that synchronizes with its perpetual calendar. MSRP: $69,900.

CHRONOMETER
GMT ±
PERPETUAL
JUN
ULYSSE NARDIN
SEP
OCT
SWISS MADE

RICHEMONT

Many of Richemont's brands straddle two categories: jewelry and specialist watchmakers, so I have included them in this list too.

Jewelry maisons: Buccellati, Cartier, Van Cleef & Arpels

Specialist watchmakers: A. Lange & Söhne, Baume & Mercier, IWC Schaffhausen, Jaeger-LeCoultre, Panerai, Piaget, Roger Dubuis, Vacheron Constantin

The A. Lange & Söhne Lange 31 Ref. 130.025 in platinum (ca. 2012). The movement is capable of an astonishing **thirty-one-day power reserve** and requires a specialized winding key to simplify the winding task.

Ferdinand Adolph Lange founded the brand in 1845 in Glashütte, Germany. The company's main production building was destroyed on the last night of the Second World War (1945), after which the area fell under Soviet control. Following the fall of the Berlin Wall (1989), F. A. Lange's great-grandson Walter joined forces with watch industry veteran Günter Blümlein to reestablish the Lange manufactory.

Today, Lange is back to its acclaimed place among the world's leading luxury watch brands. It is now part of the Richemont group of companies.

A. LANGE & SÖHNE
GLASHÜTTE I/SA
10
31
MONATS-WERK
0
MADE IN GERMANY

Brothers Louis-Victor and Célestin Baume founded Baume & Mercier in 1830. In 1918, William Baume, then company director, joined forces with Paul Mercier to found "Baume & Mercier" in Geneva. Richemont acquired Baume-et-Mercier in 1988.

Baume's current collections include the Clifton, Classima, and Hampton. Shown here is the Clifton 1892 Flying Tourbillon (Ref. M0A10454, MSRP: $59.950), first introduced in 2014 with a silver dial.

The timepiece commemorates Baume & Mercier winning the Kew Observatory timing trial in 1892 in London (with a tourbillon pocket watch). The 18K rose gold watchcase is 45.5 mm in diameter and features a mechanically wound IWC-derived movement visible through the transparent case back.

BAUME&MERCIER
GENEVE
SWISS MADE

Introduced in 2001, the Cartier Roadster (discontinued since 2012) is an attractive and sporty tonneau-shaped watch with sleek curves and racing-inspired ornamentation. Shown here is Ref. W62025V3 (last known MSRP: $7,300). Bracelet and case in stainless steel with case size measuring 44.3 by 38.7 mm. This watch pioneered a quick-release bracelet system to allow for tool-free bracelet and strap changes. The screws in the lugs (together with lug hollows) are representative of vintage roadster car lights. Ditto for the bullet-shaped crown.

CARTIER
AUTOMATIC
SWISS MADE

Jaeger-LeCoultre Ref. Q2336420 (ca. 2007, MSRP: $350,000 when new). Limited to seventy-five pieces, it is a personal favorite and an absolute marvel of horological engineering.

The JLC Gyrotourbillon 2 with platinum Reverso case houses calibre JLC 174 (371 parts, fifty-eight jewels, 28,800 vph, and fifty-hour power reserve). The dial is in silver with hobnail patterning. It measures 55 by 36 by 15.8 mm.

At six o'clock is the stunning **spherical tourbillon** with cylindrical balance spring. The inner cage rotates every 18.75 seconds. The outer cage rotates every minute. The tourbillon alone consists of one hundred parts.

JAEGER-LECOULTRE
SWISS MADE

Manual-winding Roger Dubuis RDDBEX0481 Skeleton Double Flying Tourbillon (ca. 2015, MSRP: $287,000). Black DLC titanium case. Red accents in aluminum (aluminium for British English readers). The movement consists of 301 parts and twenty-eight jewels with fifty-hour power reserve and skeletonized microrotor. The case is 45 mm in diameter.

The Poinçon de Genève hallmarking (see stamped Geneva seal insignia on dial at six o'clock) ensures superlative finishing. The twin tourbillons in calibre RD01SQ rotate every sixty seconds and connect to a differential to average out their rates. Each movement requires 1,200 hours of manufacturing. Limited to 188 pieces.

ROGER DUBUIS
SWISS
MADE

The Vacheron Constantin Jubilee 1755 in Platinum, Ref. 85250/000P. In commemoration of the brand's 250th anniversary (1755–2005), Vacheron produced a limited series of timepieces in four metals.

To the left and right of the twelve o'clock marker on the dial is Vacheron's "secret signature"; the dates 1755 and 2005 are gently engraved. Vacheron produced only 252 pieces in the 40 mm platinum case.

Calibre 2475 powers the wristwatch. The watch also features an exhibition case back and platinum pin buckle.

VACHERON CONSTANTIN
GENEVE
SAM
VEN
JEU
MER
MAR
AUTOMATIC
SWISS
MADE

Cartier introduced the "Tank à Guichets" in 1928. It quickly found its way into the hands of clientele like the Maharaja of Patiala (Maharaja Sir Bhupinder Singh, Indian royal and cricketer), one of Cartier's leading patrons.

It featured a "jumping hour" within an aperture—from whence it derives its name "guichets," which is French for counter(s) or aperture(s)—instead of a traditional analog display. This platinum homage to the Maison's signature Tank style is from 1997, when Cartier updated this industrial-looking classic in celebration of the company's 150th anniversary. One hundred and fifty individually numbered pieces were released to the public. The case measures 26 by 37 mm with the seventeen-jeweled Caliber 7001 beating inside. Cartier's deployant clasp is in 18K white gold. Current (ca. 2019) auction results in the range of $40,000–$50,000 for very fine examples.

8
40 45 50

LVMH

Watches and jewelry: Bvlgari, Chaumet, FRED, Hublot, TAG Heuer, Tiffany & Co., Zenith

Several other notable LVMH brands also produce watches but slot into other business units within the LVMH group (e.g., Christian Dior, Fendi, and Louis Vuitton).

Introduced in 2011, the TAG Heuer Mikrograph (Ref. CAR5040.FC8177, MSRP: $50,000) oscillates at an astonishing **360,000 vibrations per hour**. TAG Heuer's quest for high-frequency timekeeping began several years prior with its release of two modular one-hundredth-second chronographs in the form of the TAG Heuer Vanquish (2005) and then the Carrera Calibre 360 (2006).

This wristwatch is one of only 150 examples. Unlike its predecessors, it utilizes an integrated movement developed in-house by TAG Heuer. The main timekeeping balance vibrates at an ordinary 28,800 vibrations per hour, while the dedicated chronograph balance runs 12.5 times faster. The blue one-hundredth-second hand races around the dial once per second while the chronograph is active.

100
MIKROGRAPH
HEUER
SWISS MADE

Ref. CAR5A10.FT6034, the “Mikrotimer Flying 1000 Chronograph,” raised the frequency bar yet again for TAG Heuer. Only eleven examples exist of the Mikrotimer Flying 1000 (MSRP: $88,580—by invitation only). With a frequency ten times that of the Mikrograph, the mesmerizing yellow hand flies around the dial once per second at **an insane 500 Hz or 3.6 million beats per hour** (500 × 2 × 60 × 60).

The bottom subdial records up to five seconds. Every revolution of this lower dial (i.e., five seconds) requires the advancement of a white hand (hidden below the yellow chronograph second hand and just barely visible in the image). After one minute, the white hand will rest at the 60 marker in the inner dial (at the five o’clock position—see the large “1” positioned just above the 60). TAG Heuer followed it up with the Mikrogirder 2000 and 10000 (the same model, although the latter was to be the production piece; **both operated at an incomprehensible 7.2 million beats per hour**; MSRP: €125,000, ca. 2013). I have yet to see a production Mikrogirder in the wild.

1000
MIKROTIMER
5 sec / rev
SEC
SWISS MADE

Bvlgari (pronounced Bull-gah-ree) Octo Finissimo Chronograph GMT Automatic (Ref. 103068, MSRP: $17,600, ca. 2019). 42 mm diameter, sandblasted titanium case and bracelet.

Exhibition case back. GMT subdial at three o'clock advanced by a corrector button at nine o'clock. In-house Bvlgari calibre BVL 318 (3.3 mm thick, 37.2 mm diameter), powered by a peripheral platinum rotor. The Octo Finissimo collection with its record-setting thin cases has proven very popular with collectors and reviewers alike.

BVLGARI
SWISS MADE

Hublot Big Bang Unico Sapphire (Ref. 411.JX.4802.RT, MSRP: $57,900, ca. 2016). The strap is translucent to match the sapphire crystal case's translucent/transparent effects.

The movement is calibre HUB 1242 UNICO, with column wheel and flyback chronograph. Power reserve is seventy-two hours.

HUBLOT
SWISS MADE

Zenith Elite 6150 Ref. 03.2272.6150/51.C700 (MSRP: $6,700, ca. 2016) with in-house Zenith Elite calibre incorporating one hundred hours of power reserve with its double-barrel construction.

The 195-component movement is just 3.92 mm thick, making the overall watch extremely slim and elegant (42 by 10 mm). The blue sunburst dial of this everyday dress watch displays a lustrous sheen that changes with the position and intensity of the light source.

ZENITH
SWISS MADE

CITYCHAMP WATCH & JEWELLERY GROUP LIMITED

Citychamp Watch & Jewellery Group Limited (formerly known as China Haidian Holdings Limited) is a Hong Kong–listed publicly traded company. The group owns Corum, Dreyfuss & Co, EBOHR, Eterna, J&T Windmills, Rossini, and Rotary.

From one of Corum's most popular collections, the Admiral 45 Chronograph. For many years, Corum sponsored several international regattas. The Admiral's Cup was probably its most famous affiliation, lasting more than two decades.

Shown here is Ref. A116/03475 (ca. 2018, MSRP: $10,900) in prepatinated bronze with teak dial (resembling the deck of a historical tall ship) and distressed calfskin leather strap. The twelve pennants on the dial are traditional markers for this collection.

CORUM
ADMIRAL

ROLEX

The undisputed king of watch brands, Rolex is an industry and ecosystem unto itself. Rolex watches have by far the most active secondary market of any luxury watch brand. Rolex is also inarguably the most widely recognized global luxury watch brand. Not only is it the most renowned *watch* brand, it is also one of the most renowned *brands*—period.[247] Its name and distinctive crown logo are instantly recognizable, and many consider Rolex to be the ultimate symbol of status and good taste.

Some authors and historians claim that the name itself is pure fabrication, like Kodak.[248] Some claim that it is a portmanteau[249] of "ho**Rol**ogical **Ex**cellence." Still others hold that the word "Rolex" is onomatopoeic; it is the sound of a watch undergoing winding. The word for watch in Greek is ρολόι (transliterated as *rolói*), which, when pronounced, sounds like ROY-LOY. It is a very short linguistic hop to ROLEX from ROLOI. However, despite all of these hypotheticals, Hans Wilsdorf (Rolex founder) himself, in a talk delivered on July 2, 1958, stated, "I tried combining the letters of the alphabet in every possible way. This gave me some hundred names, but none of them felt quite right. One morning, while riding on the upper deck of a horse-drawn omnibus along Cheapside in the City of London, a genie whispered 'Rolex' in my ear."[250] The Rolex watch group, officially Rolex SA, has operated for more than a century. Its wholly owned subsidiary Montres Tudor SA (selling under the brand name Tudor) contributes to an overall estimated group revenue of $5.1 billion and an annual production of 800,000 watches.

The Rolex Cosmograph Daytona 116500 (ca. 2016) with black dial and (updated) black ceramic bezel. Arguably the most sought-after new Rolex (along with the Submariner), if not **the most sought-after new luxury watch, period**. This is a grail watch for many. Rolex utilizes specialized stainless steel (904L) for enhanced corrosion resistance. They call this Oystersteel. Housed within the case is calibre 4130 (an in-house movement that is also a certified chronometer). While the current retail price is $13,150 (2020), since this timepiece has a waiting list of several years, new in-box models sell in the secondary market for around $35,000, more than double the retail price.

247. According to *Forbes* as of May 22, 2019. #78: World's Most Valuable Brands 2019.

248. The April 1962 issue of *Kiplinger* magazine cites George Eastman: "A trademark should be short, vigorous and incapable of being misspelled to an extent that will destroy its identity. It must mean nothing. The letter K had been a favorite with me—it seemed a strong, incisive sort of letter. Therefore, the word I wanted had to start with 'K.' Then it became a question of trying out a great number of combinations of letters that made words starting and ending with 'K.' The word 'Kodak' is the result. It became the distinctive word for our products."

249. Truly one of my favorite words, a portmanteau is a linguistic blend of words in which multiple words or their sounds (phonemes) are combined into a new word (e.g., hebra, from horse and zebra, or zorse from zebra and horse—both of which, surprisingly to some, are real creatures).

250. Rolex.org, a division of Rolex SA.

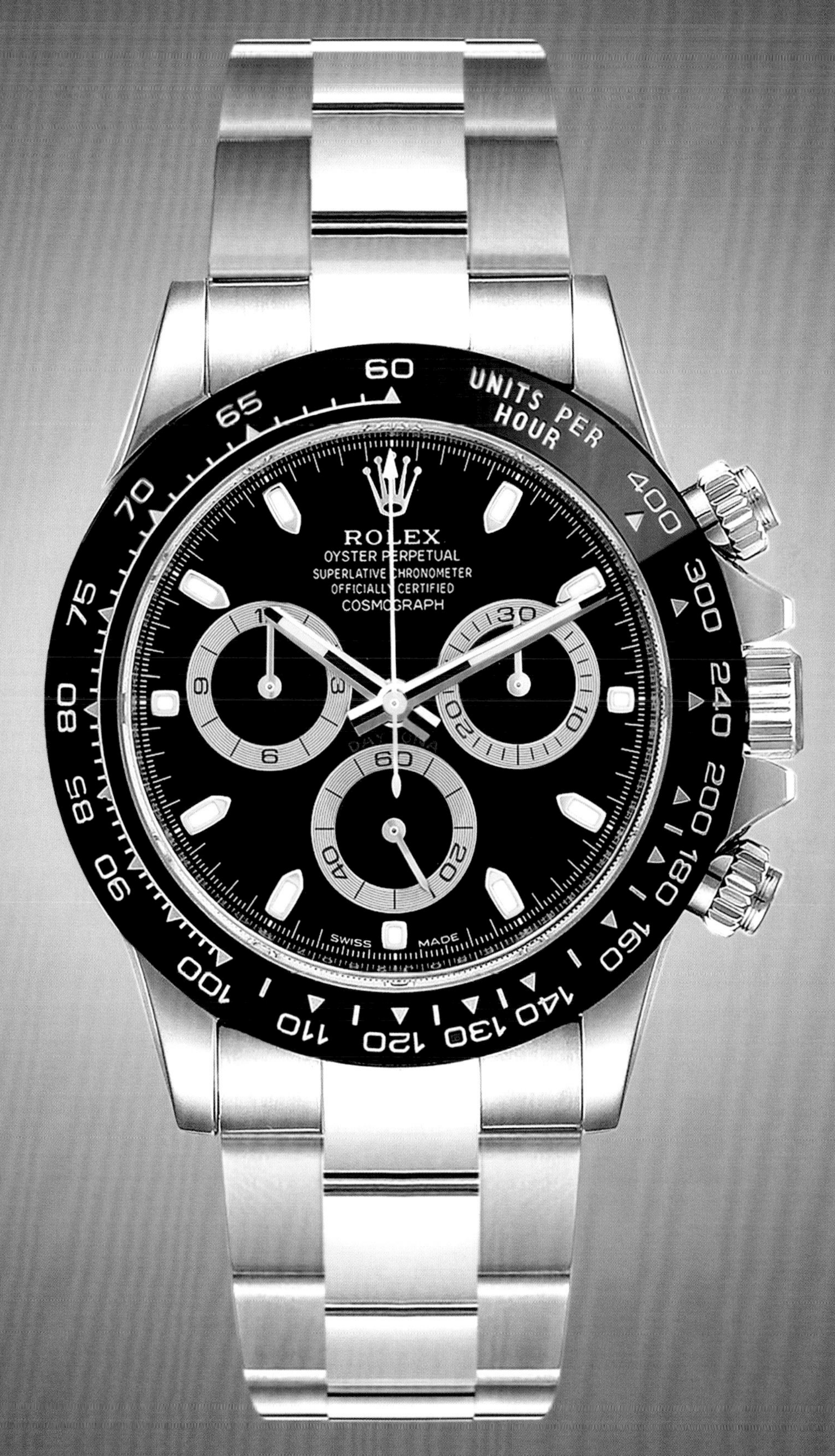
UNITS PER
HOUR
ROLEX
OYSTER PERPETUAL
SUPERLATIVE CHRONOMETER
OFFICIALLY CERTIFIED
COSMOGRAPH
SWISS
MADE

THE BIG INDEPENDENT BRANDS

AUDEMARS PIGUET

The brand Audemars Piguet (AP) is almost synonymous with its most famous creation, the Royal Oak (first presented at Baselworld in 1972 and considered the first luxury sports watch), and its derivate, the Royal Oak Offshore (released in commemoration of the twentieth anniversary of the Royal Oak). Please refer to the chapter on watch design for further information about the Royal Oak and its design development.

Founded by Jules Louis Audemars and Edward Auguste Piguet in 1875, the company became Audemars Piguet & Cie in 1881. It has been family owned since formation.

In September 2018, CEO Francois-Henry Bennahmias declared to Reuters, "Since the beginning of the year, we've seen double-digit sales growth. We were already close to one billion Swiss francs in sales last year; we'll easily exceed it." With 40,000 units[251] produced annually, AP commands a roughly $25,000 average selling price at wholesale—a very high average price per unit. Since AP intends on opening additional stand-alone stores and reducing the number of cobranded outlets, higher margins associated with DTC merchandising will only increase the brand's profitability and revenue values.

Audemars Piguet Royal Oak Offshore Grand Prix (Ref. 26290IO.OO.A001VE.01, ca. 2010). Hand-stitched calf leather with titanium and Alcantara inserts.

Forged carbon case. Exhibition case back. Limited to 1,750 pieces. Manufacture Calibre 3126/3840 (365 parts). Fifty-hour power reserve. (MSRP: $42,700)

251. Bloomberg News, January 16, 2018.

TACHYMETRE BASE 1000
AUDEMARS PIGUET
SWISS
MADE

BREITLING

Léon Breitling founded the eponymous watch manufacturer in 1884 at the relatively young age of twenty-four. By the late 1990s, the brand had become the global standard-bearer for pilots and navigators. As a pioneer in aviation timekeeping, Breitling has continued to produce wristwatches of distinction. Breitling's renowned Chronomat, introduced in 1942, was the first watch to incorporate a circular slide rule. A Breitling timepiece is instantly recognizable for its large case and distinctive styling. In 2009, Breitling introduced its own in-house calibre, the B01. Traditionally, Breitling had sourced its movements from Valjoux, ETA, and Venus.

Breitling's most recognized collections include Navitimer, Avenger, Aerospace, and Emergency. In 2002, Breitling began a collaboration with the luxury automotive brand Bentley to produce watches that incorporated distinct design motifs from both brands. Rumor has it that Breitling's annual production is between 200,000 and 250,000 units, with annual revenue approaching $600 million. Since it is a privately held company, precise numbers are not publicly available.

PATEK PHILIPPE

Watchmaker Antoni Patek started making pocket watches in 1839 in Geneva, along with his fellow Polish migrant Franciszek Czapek. They separated in 1844, and in 1845 Patek joined with the French watchmaker Adrien Philippe, the inventor of the keyless winding mechanism. Patek Philippe & Co incorporated in 1851.

Patek Philippe pioneered the perpetual calendar, split-seconds hand, chronograph, and minute repeater in wristwatches.

Like other Swiss manufacturers, the company produces mostly mechanical movements of the automatic- and manual-wind varieties but has produced quartz watches in the past, and even a digital wristwatch, reference 3414.

Patek Philippe is notable for manufacturing its own watch components. The Stern family has owned Patek Philippe since 1932, and Thierry Stern currently helms the operation.

Patek offers more than 170 models within several watch collections, including Calatrava, Twenty4, Golden Ellipse, Gondolo, Aquanaut, Nautilus, Complications, and Grand Complications (which feature celestial charts, minute repeaters, and perpetual calendars).

Rolex may be the undisputed king of watches, but if the watch world had an elder statesman, it would undeniably be Patek Philippe.

No other brand garners as much attention as "Patek."[252] Its position in watchmaking is enviable. It produces roughly 60,000[253] timepieces annually and generates a not-inconsiderable CHF 1.5 billion. Moreover, demand outstrips supply many times over. Informed collectors are well aware of supply dynamics, and no other brand save for Rolex (and perhaps select Audemars Piguet and Richard Mille offerings) commands such outsized premiums—not just for collectible vintage timepieces, but also for brand-new timepieces.

The Patek Philippe 5004R-014 (ca. 1994-2012). A rare wristwatch with just over two hundred known examples.

The 5004, introduced under Philippe Stern's brand stewardship, is a highly collectible timepiece. Auction prices of this watch are still undervalued. Expect prices to strengthen considerably when collectors recognize its worth (last known MSRP: $236,500). Its collectability is enhanced by the fact that the split-seconds movement incorporates an ingenious isolator mechanism referred to by collectors as the "Octopus" (because of its tentacled isolator wheel, albeit six pronged).

The watch measures 36.7 mm in diameter despite housing numerous complications in calibre CHR27-70Q. Among its complications: perpetual calendar (leap year, month, day, and date), twenty-four-hour dial, and split-seconds chronograph. The visible apertures at ten o'clock, 11:30, 12:30, and 6:30 adjust the complications.

252. Wristwatch cognoscenti refer to Patek Philippe simply as Patek.
253. Estimates put this number for 2020 at three-quarters of 2019's output.

SAT
JAN
PATEK PHILIPPE
GENEVE

Sought-after Patek complications sell for many times their list price at first release. Even relatively "mainstream" steel watches are allocated sparingly to the most-prominent authorized dealers and customers, and it is not uncommon to find these watches in the gray market selling for three or four times list price.

As of early 2021, the Patek Philippe Nautilus 5711/1A-010 (introduced in 2006 on the thirtieth anniversary of the Nautilus) is selling for an average of $95,000 through the secondary market yet has a sticker price of *only* $30,620. How long this asset bubble will last is anyone's guess, but sustained interest from Asia and astronomical auction prices of vintage examples have helped keep these values aloft.

RICHARD MILLE

Richard Mille is the (relatively) new kid on the block, founded just before the turn of the millennium (1999). With few exceptions, no other watch manufacturer departs so far and so frequently from the traditional techniques of watchmaking.

Unveiled to the world in 2001, the RM 001 Tourbillon broke from watchmaking tradition. From its avant-garde design to its innovative use of materials and manufacturing processes, the RM 001 astonished the watch world. Now well into the RM series, Richard Mille continues to produce racing-inspired timepieces of unique design.

Richard Mille, the man and founder of the eponymous brand, lives the jet-set life the brand exudes. He is also an avid vintage race car collector and driving enthusiast.

With starting prices now around $80,000, this brand makes several watches that sell new for more than $1 million. In a number of interviews in 2019, Mille said that the brand's target was to produce 5,200 watches for the year (up from 2018's 4,600 watches).

Richard Mille watches sell at a price point that beggars belief, and—if we were discussing automobiles—are akin to Koenigsegg, a Swedish manufacturer of high-performance multimillion-dollar luxury hypercars. The mere fact that one can afford such a car/timepiece is reason enough for its acquisition; that they are also archetypes of manufacturing excellence may seem beside the point, but it is in fact part of the charm of the brand. The level of fit and finish is exquisite. The use of new materials is also unprecedented. While others have subsequently copied, Mille was the first to make a watchcase entirely out of sapphire crystal.

Mille's watches are no delicate trinkets either. Rafael Nadal is famous for wearing a tourbillon while playing championship-level tennis. The RM 27-03 weighs a scant 31.7 grams and yet is capable of withstanding shocks of up to 10,000 g. With annual revenues approaching CHF 1 billion, Richard Mille commands the highest average price per watch of any brand (other than perhaps the most elite boutique microbrands). As a show of status alone, these Veblen goods[254] are peerless.

The RM005 from Richard Mille (ca. 2004, MSRP: $35,000). The most affordable Richard Mille timepiece at the time, it was subsequently replaced with the larger RM010 (launched in 2006). The movement comes from Vaucher SA. The tonneau-shaped titanium case measures 37.8 by 45 mm.

254. Named for American economist Thorstein Veblen. He first identified conspicuous consumption as a way of seeking status, as outlined in his book *The Theory of the Leisure Class: An Economic Study of Institutions* (1899). Veblen goods are luxury goods for which the quantity demanded INCREASES as the price increases. This seems to contradict our normal understanding of how demand operates. The result is an upward-sloping demand curve. Veblen goods may be positional goods (i.e., something few others can own; an item desired because of its high price).

RICHARD MILLE
RM005
SWISS MADE

CHOPARD

From pocket watches to jewelry watches, Chopard has been manufacturing reliable timepieces since 1860. It was then that Louis-Ulysse Chopard[255] set up his Swiss Jura-based factory. After the business was in the Chopard family for three generations, the Scheufele family acquired it in 1963. Their hands-on ownership has transformed the prestigious brand. Chopard is a longtime sponsor (going on more than thirty years) of the Mille Miglia, a 1,000-mile Italian road race, and has a line of timepieces named for the race. These racing-inspired timepieces are quite collectible. In fact, more than a few aficionados have every edition in their collection. Chopard's estimated total annual revenue is about $800 million, and the company produces approximately 75,000 timepieces per year.

Chopard Mille Miglia Chronograph Model 8331, ca. 1999/2000 (original MSRP: $2,600). Silvered circular-grained dial with running seconds, thirty-minute counter at nine o'clock, and twelve-hour counter at six o'clock.

Tire-patterned (Dunlop tread) rubber strap is a hallmark of Chopard's Mille Miglia designs, although several are also available with leather straps (or bracelets). Limited to a thousand pieces in steel with exhibition case back.

255. Chopard's sub-brand is named L.U.C. in honor of its founder.

TACHYMETRE
Chopard
GENEVE
AUTOMATIC
MINUTES
HOURS
1000 MIGLIA
1999
SWISS MADE

HERMÈS

Founded in 1801 as a saddle and harness maker, Hermès is now a global brand with almost €6.9 billion in annual revenue.[256] In 2017, following acquisitions of Natéber SA and Joseph Erard SA, Hermès formed "Les Ateliers d'Hermès Horloger" as its watchmaking unit. The Arceau collection, first introduced in 1978, sits alongside the Cape Cod, Faubourg, and Heure H collections. Watches account for only €193 million at wholesale cost.[257] In 2015, Hermès collaborated with Apple Inc. to produce specially crafted watchbands for the Apple smartwatch.

SEIKO

In 1881, twenty-one-year-old Kintarō Hattori opened "K. Hattori" in the Kyobashi district—today part of the Ginza district—in Tokyo. He built and repaired watches, clocks, and jewelry. A decade later, Hattori began producing his own clocks under the brand name Seikosha ("House of Exquisite Workmanship"). In 1924, after the Great Kanto earthquake leveled Seikosha's factory, Hattori rebuilt and renamed the company Seiko, meaning "exquisite" or "success." He released a 24 mm, subsecond, three-hand wristwatch in the same year, with "Seiko" adorning the dial for the first time. Seiko released its first in-house movement in 1956 (housed in the Seiko Marvel) and followed up with the first Grand Seiko wristwatch in 1960. Twenty-first-century Grand Seiko designs still reference the original motifs: sharply chamfered dauphine hour and minute hands; mirror-polished hands, markers, and case; and distinctive chamfered lugs. Watch cognoscenti consider Grand Seiko finishing on a par with even the most storied of Swiss brands. Some consider it unrivaled.

The Seiko Holdings Corporation is today a billion-dollar entity selling watches, clocks, electronic devices, semiconductors, jewelry, and optical products. Within the watchmaking division, Seiko has a multitude of brands—entry-level Seiko 5, luxury brands Credor and Grand Seiko, and SARB, Prospex, and Presage. Seiko also produces a movement called Spring Drive, announced in 1997 and accurate to within one second per day. The movement uses a traditional mainspring and gear train as energy source and transmission mechanism, but it replaces the escapement and balance wheel with a Tri-synchro regulator designed to act as a mechanical brake, operate at lower noise levels, and provide a continuous gliding motion to the second hand.

Seiko's annual revenue for wristwatches exceeds $1.25 billion.[258]

Another personal favorite, the Grand Seiko Ref. SBGH277 in stainless steel, featuring calibre 9S85 (MSRP: $6,100). A 40 mm case with curved sapphire crystal on front and back.

Grand Seiko's Mechanical "Hi-Beat" movement oscillates at 36,000 vibrations per hour (vph). The entire watch is highly polished with sharp delineations between surfaces, creating a harmonious interplay between light and shadow.

Even the hands are polished to the extreme, using a technique called "Zaratsu" (similar but not identical to Swiss black/mirror polishing). This technique involves polishing surfaces to a distortion-free mirror finish—a hallmark of all Grand Seiko timepieces, making them instantly recognizable.

256. 2019 Hermès Group annual report.

257. Ibid.

258. Seiko consolidated financial statements (fiscal year ending March 31, 2020). Seiko Holdings' total revenue for the same period was JP¥ 239,150 million (about $2.2 billion).

GS
Grand Seiko
AUTOMATIC
HI-BEAT 36000
4

CITIZEN

The Citizen Group came together in 1918 "with the dream of producing a watch made in Japan."[259] Approximately the same size as the Seiko Group, Citizen is Japan's other watchmaking powerhouse. In addition to Citizen's in-house brands, the Tokyo-based company includes Bulova, Frederique Constant, La Joux-Perret, Arnold & Son, and movement maker Miyota within its brand portfolio. Approximately 51 percent of the holding company's revenue derives from watches, constituting $1.7 billion.[260]

FREDERIQUE CONSTANT

The first modern "affordable" all-Swiss brand, Frederique Constant positioned itself as a brand with classically inspired (although not particularly innovative) designs for far less than the higher-priced established brands. Acquired in 2016, it is now a subsidiary of Citizen. Husband-and-wife entrepreneurs Peter and Aletta Stas established it in 1988, and its popularity has helped position the brand firmly within the "accessible luxury" segment: CHF 1,000–5,000. The company produces over 100,000 timepieces annually and owns two other brands, Ateliers deMonaco and Alpina, helping the company bestride price points above and below that of Frederique Constant.

CASIO

Casio Computer Co., Ltd., headquartered in Shibuya, Tokyo, produces calculators, mobile phones, digital cameras, electronic musical instruments, and analog and digital watches. Tadao Kashio established Casio as "Kashio Seisakujo" in April 1946. From humble beginnings, Casio produced its first calculator prototype in December 1954.[261]

After successfully developing and launching the Casio Mini calculator (in 1972, with first-year sales eclipsing one million units), by the early 1970s Casio was ready to diversify into timepieces. Utilizing technology developed for electronic calculators, it was a natural extension for the company.

In October 1974, Casio introduced the CASIOTRON. It displayed not only hours, minutes, and seconds, but also a.m./p.m. indication, month, date, and day of the week. The first watch with a Casio-produced LCD launched in September 1978.

April 1983 saw the introduction of Casio's legendary shock-resistant watch in the form of the G-Shock DW 5000C. Casio's March 2020 year-end financial results show net revenue of approximately $2.6 billion. Casio does not break out financial numbers by industry segment, so it is difficult to know what contribution timepieces make to overall sales, but on the basis of the popularity of the iconic G-Shock and other Casio watches, it is probably substantial.

Casio TC-500 eight-digit capacitive touchscreen calculator-watch (ca. 1983, original MSRP: $79.95; case in chrome, steel, and resin). As a teenager, I was gifted a TC-500 for my birthday, and other than a battery change, it required no servicing. I wore it continuously for almost eight years. Casio did not invent the calculator watch—Pulsar (1975, Ref. 1822 for $3,950 in 18K gold) and HP (1977, Ref. HP-01 for $450–$800) were first to introduce them (with red LEDs)—but no brand has been more prolific or celebrated than Casio in its calculator watch offerings. This watch is still one of my favorites.

In addition to the obvious calculator functionality—including powers and reciprocal functions—performed using the touchscreen, the watch also features a day/night indicator, backlight, three independent alarms, chime function (on the hour), dual time, twelve-hour countdown timer, one-hundredth-second chronograph (with split times—accumulating up to twenty-four hours), and perpetual calendar. Accurate to within fifteen seconds per month, it features a capable lithium battery. The attached bracelet, as with those for most Casio watches of the period, appears to have been designed for Lilliputians.[262]

259. Citizen Annual Report 2019.
260. Ibid.
261. Casio, https://world.casio.com/corporate/history.
262. Lilliputians are an imaginary society of tiny people about 6 inches in average height. Described by Jonathan Swift in his satirical fantasy novel *Gulliver's Travels* (1726), these fictional characters, despite being tiny in stature, exhibit a sense of self-importance associated with human adult men.

CASIO
TC-500
TOUCH SENSOR CAL
12345678.
ADJUST
MODE
LIGHT
12•24HR
[7] [8] [9] [0]
[4] [5] [6] [•]
[1] [2] [3] [=]
C
÷ × − +

These images present the Jaeger-LeCoultre Reverso Squadra World Chronograph (Ref. Q702T470, last known MSRP: $19,600), first offered for sale in 2006 and another one of my all-time favorite watches (I have quite a few favorites). The Reverso Squadra took an already winning formula and updated it for the twenty-first century—although according to lore, the squarish design, as opposed to the more traditional and recognizable rectangular Reverso, ended up being sidelined at design inception due to its rather modern proportioning. Too avant-garde for the time, it had to wait seventy-five years before being unveiled to the public.

Originally created for the wrists of British army officers stationed in India, Jaeger-LeCoultre conceived and built the Reverso to withstand the rigors of the sport of polo, where a misdirected mallet or ball could obliterate a standard watch crystal. The ingeniously designed timepiece originally displayed a case back of pure metal—completely unadorned but for an occasional customized engraving. The reversible case swung around with ease to protect the delicate crystal, dial, and other internals from a devastating blow. During the polo match, the case swung around effortlessly, revealing the dial. After reading off the time, the wearer returned it to its facedown position and to safekeeping for the rest of the match.

Registered in 1931, the patented design is still recognizable today in modern editions of the Reverso. The horizontal gadroons[263] above and below the case opening and just below the lugs are emblematic of this legendary design. For almost a century, at least some version of the Reverso has been a mainstay in Jaeger-LeCoultre's collective offerings. The 41 by 35 by 13 mm case (53 mm lug to lug) is fashioned from Grade 5 brushed and polished titanium—making it the first-ever Reverso manufactured in this hard-to-work metal. It features 50 meters of water resistance and a sixty-five-hour power reserve and is limited to 1,500 numbered examples. In the rubberized-metal bracelet, each link is individually removable—a welcome addition for those with small wrists.

This beautiful and iconic timepiece features a simultaneous twenty-four-zone world-time display on the reverse that is set on demand via the crown. A transparent disc above the world's cities displays a day/night line of demarcation. The dial is modern, with the square theme carried over to the subdials and the big date at twelve o'clock. The horizontal gadroon aesthetic parallels the central dial texturing. The running seconds at six o'clock overlay a red/white day/night indicator in the upper half of the quadrilateral.

The hand-assembled and hand-decorated movement is the unidirectional-winding mechanical automatic JLC 753 with ceramic rotor bearings. It beats at 28,800 vph, features 366 parts and thirty-nine jewels, and is 7.57 mm thick.

263. A decorative edging on metal or wood, typically formed by inverted flutings.

CHRONOGRAPHE
SWISS MADE

CHRONOGRAPHE
WORLDTIME
AUTOMATIQUE
SWISS MADE
CHICAGO
DENVER
LOS ANGELES
ANCHORAGE
HONOLULU
SAMOA
AUCKLAND
NOUMEA
SYDNEY
TOKYO
HONG KONG
BANGKOK
DHAKA
KARACHI
ABU DHABI
MOSCOW
CAIRO
PARIS
LONDON GMT
AZORES
H2O
R.JANEIRO
CARACAS
NEW-YORK

Casio's G-Shock line is the flagship collection with a very active collector community. Other collections include Edifice, Wave Ceptor, Data Bank, Pro-Trek, and Classic (digital or analog).

To commemorate thirty-five years of G-Shock, in 2019 Casio released just thirty-five numbered examples of reference G-D5000-9JR—in solid 18K gold. It took three years to come to market because of technical difficulties in shock proofing and waterproofing the all-gold watch to 200 meters. Priced at 7.7 million yen, or about $70,000, it is the most expensive G-Shock ever made.

FINAL NOTE

"Happy were they who stood in that place in that hour."

–Rabbi Eliezer, son of Hyrkanos

Dear reader, I hope you have enjoyed reading this book as much as I have enjoyed writing and illustrating it. Modern mechanical watchmaking is one of the last vestiges of human prowess in fashioning fine micromechanical machinery. As automotive, aviation, appliance, and portable consumer technology become ever more automated and laden with depersonalized digital circuitry, we as a species are losing touch with the artisanal crafts.

Many of the images shown in this book are of watches well beyond the reach of most of us in terms of rarity and price. Buying a mechanical watch does not need to break the bank. Inexpensive, good-quality mechanical watches are available for well below $1,000—many even below $500. Preowned good-quality watches are less expensive still.

Smartwatch technology continues to become commonplace, but a mechanical timepiece with its tick-tick-tick analog timekeeping continues to connect us with our past and helps us maintain a more intimate relationship with time.

This book went to publication during the devastating COVID-19 pandemic. The worldwide virus has taken a heavy toll on humanity. It has also taken a considerable toll on the global economy. The traditional watch industry has been affected perhaps more so than most. Luxury is unnecessary during a crisis. Will the watch industry recover? It is too early to tell, but 2020's financial results already show historic declines. Many watch companies, weakened by slowing demand brought about by the smartwatch revolution and changing consumer tastes, will end up shuttering permanently. The watch industry has had its challenges, but this may be its biggest test yet.

EPILOGUE

I began to write this epilogue after sending my final manuscript to the publisher on March 25, 2021. I will have little input for the next twelve (or more) months. I've decided that my publisher-edited manuscript should have no further edits but for grammar and spelling and the personalized editorial choices of the amazing Schiffer team. One reason is that it's just going to be too difficult to update all the industry numerical and financial data with so little expected review time between final edit and publication date. I'm also not sure when I'll be called upon to make the edits.

I have updated these additional thoughts from late March through July 2021. Unfortunately, my dear father, of blessed memory, passed from this world last month (June 14, 2021), and I wish now to focus on my family, not on updating this book, so this is my epilogue.

The current publication date is scheduled for around the first quarter of 2022, but I wouldn't hold my breath given the uncertainties of supply-chain disruptions during COVID. And, if there is a secondary world-changing event during all of this—several adversarial states still look increasingly menacing (and pandemics often portend propitious timing for belligerent state antagonists looking to capitalize on stretched resources)—my predictions about the publication date could take a further hit.

The 2019–2020 numbers and values I have shown together with my analysis of where I think things are going should prove instructive two years hence. So I'm leaving everything I previously wrote completely untouched, and in a few paragraphs, I'm going to do some forecasting, just for fun. Let's see how far off the mark I am when 2023 rolls around about eighteen months from now.

Rolex, by the end of 2022, may command more than 30 percent of luxury watch industry revenue, at around $9 billion, from the sale of more than 1.15 million timepieces. Here's why: Because their ecosystem is so well insulated from the effects of the coronavirus (they plan for situations like this and other unexpected global disasters well in advance), even if the virus mutates through the entire Greek alphabet (excepting global war, which would change everything written here), the huge swathes of wealth accumulation through government handouts, a fairly hot stock market (which could change very suddenly), newly minted crypto millionaires, and accelerated personal saving will concentrate immense amounts of wealth with very few productive investment-asset choices, at least on appreciating durable goods.

Rolex is nothing if not consistent. Rolex will likely increase inventory availability by some several hundred thousand units, but no more, because, well, they're Rolex and they can (relatively speaking) smooth supply bottlenecks as they choose and keep demand at full tilt.

Unlike past asset bubbles, expect almost every Rolex reference to become "desirable," including the normally off-the-shelf, available-to-purchase-immediately Datejust, and look for these mainstays to sell at or above MSRP as people turn to the secondary market for a quick profit on their cherished authorized dealer allocations.

This dire situation will be especially acute for Daytonas and Submariners, which will prove very, very hard to get even at "reasonably ludicrous" prices, with no end to these aftermarket premiums in sight. Secondary-market prices of Patek's Nautilus and those of similar ilk will go absolutely bonkers (a well-established but rarely used industry term meaning "very high").

And should Patek come out with another Nautilus 5711, the final-final-final edition will go through the stratosphere in secondary-market value. I expect Patek may even pick a super-exotic, trend-setting color. Last year's Pantone color of the year was blue, but they already have a blue, so who knows? Patek is known to do exotic dials at the end of a model era. Remember Salmon dials a decade ago?

This is going to be a rapidly inflating economic environment for at least the next two or more years, possibly followed by a recession, and the US Federal Open Market Committee's monetary and interest rate policy is going to come under incredible strain and critique.

I expect mortgage rates to spike above 6.5 percent and capital to flow into appreciable fixed asset investment classes before rates get too high. Houses, cars (especially rare naturally aspirated/manual shift varieties), gold, platinum, palladium, lithium, watches, and similarly positioned (collectible) assets will be in huge demand with a major supply shortage brought about by lingering COVID-19 effects (and super-rapid electric vehicle adoption). And don't forget "maximalism," which I talked about earlier in the book. Let's also not overlook the auction market. Phillips, Christie's, Bonhams, Sotheby's, and others are going to have banner years, and the vintage and preowned market also will grow rapidly as more entrants get into luxury watchmaking.

Collectible watches, especially Rolexes, may as well serve as a reserve currency, so bulletproof will they appear as an inflationary hedge (and their very own micro asset-class), for the short-term at the very least. The rest of the remaining industry participants (some brands will unfortunately and disappointingly disappear permanently) in the above 3,000 CHF price point will flourish, and I expect that boutique brands like Grönefeld and Hajime Asaoka (and similar low-volume artisanal manufacturers) will not be able to keep up with even half of their enormous expected demand.

Not so for the sub-$500 price point, which will continue to decline in total export value (as smartwatches ascend above 100 million annual units sold towards the end of 2022—or even earlier). Brands like Swatch and other lower-priced timepieces will need to massively leverage group strength. Maybe Swatch teams up with Longines, Tissot, or—dare I say it—the crown jewel Omega (super risky but genius if they can make it work), one of the three heavy hitters for the group, to drum up additional demand in young, fashion-savvy customers. Expect Grand Seiko to take advantage of the extreme supply shortages. Grand Seiko, when introduced to the US in 2010, was my number one choice for most improved a decade later, maybe excepting Richard Mille. (Let's be honest; the original RM005 and RM010 were early favorites of Mille's avant-garde sculptures.) If memory serves I even sought out an authorized Grand Seiko dealership in 2010. I was rejected; no hard feelings, Seiko. I expect even better results from Grand Seiko by the end of 2022.

Watch dial colors are going to proliferate. You're going to see almost every brand come out with exotic colors to shed the dour five-plus years of pandemic outlook (this is how long they historically last). But don't get carried away; too many "funky" references are how supply gluts start. Soon enough, "boring" old black, white, and silver will be roaring back. And green will be even more popular as a dial color by 2022's end. I have already written about past design reversals as a sociocultural reaction to the end of periods of lowered public expectations and confidence. Be warned: watch industry watchers should take care to monitor the growing inflated watch asset bubble for signs of a near-term pop, and lesser manufacturers should maybe produce (a little) less than total expected demand, not the other way around, to prevent another inventory glut.

Based on current trajectories and management strength, as it stands today, revenue leaders at the end of 2022 will likely be Rolex, Cartier, Omega, Audemars Piguet, Richard Mille, Patek Philippe, and Longines. Also expect a good showing from Breitling when their updated management plans (under new private ownership) fully kick in. Bvlgari will continue to delight.

How will the industry look beyond 2022? Perhaps, if demanded, I'll get to write a follow-up book.

Be healthy. Be safe. Thank you so much for reading.

Barry B. Kaplan
Wednesday, July 21, 2021

ABOUT THE AUTHOR

"Time is an illusion."

—Albert Einstein

Barry Kaplan is a third-generation watch industry contributor. In his spare time, he enjoys carpentry, sculpture, and design. He also enjoys the world of automobiles and is almost as obsessive about cars[264] and motor racing as he is about wristwatches.

Barry has worked in the watch industry for almost two decades and has consulted to various organizations and industry groups regarding strategic and operational management, product development, industry trends, and the rise of the smartwatch.

Barry holds degrees in accounting, business finance, and corporate administration. He continues to work and consult in the luxury and wristwatch industry. He also consults to companies in the burgeoning augmented reality (AR) space regarding gestural control and associated user interface design.

Photo by David Zimand

264. According to Barry's parents, his first spoken word was "car," even prior to voicing "Mom" and "Dad."

ACKNOWLEDGMENTS

"When we give cheerfully and accept gratefully, everyone is blessed."

—Maya Angelou

This book would not have been possible without the contributions, both large and small, of many individuals.

A special thank-you to my wife, Emily, for her unrivaled support and encouragement, for putting up with me while writing and illustrating this tome, and for offering numerous editorial insights and creative ideas.

Thank you to my kind and generous parents and grandparents for fostering within me inquiring-mindedness and a love of watchmaking.

Thank you to my amazing children for keeping me young, if only in attitude, and for your contributions to various segments of the book.

A big thank-you to the generous watch collectors for letting me photograph your treasures, some going back more than a decade.

Thank you to Nicholas Manousos, executive director at Horological Society of New York, for reviewing several technical matters regarding the watch movement.

Thank you to Rabbi Yaakov Zev Smith for your thoughtful contribution, especially where it concerns biblical authenticity and accuracy.

Thank you to Rabbi Larry Rothwachs for your perceptive halaḥic guidance.

Thank you to Rabbi Dr. Akiva Tatz for your immeasurable inspiration.

Thank you to Rabbi David Shapero for your novel approach to the twenty-four-hour day.

Thank you to Ze'ev Atlas, Yale Butler, Myron Chaitovsky, Mitchell First, Maya Balakirsky Katz, Terence Klingman, Moshe Rosenberg, Menachem Shapiro, and Elnatan Sulimanoff for your kind guidance regarding theological subject matter and its appropriate translation.

Thank you to engineer and educator Arvin Ash for your generous comments regarding entropy.

I would particularly like to highlight the very generous involvement of Rabbi Dr. Richard Schiffmiller, especially in the area of physics. I am extremely grateful for his remarkably thorough scholarly perusal of the chapters on "The Measurement of Time," "The Philosophy and Science of Time (and Space)," "Time Travel," and the appendix.

Thank you to David Mansell and Richard Mansell for your insightful counsel.

Since I do not wish to change the original manuscript one iota, except for editorial decisions taken by Schiffer Publishing, I would personally like to use this paragraph to thank them for their efforts. It hasn't always been easy (during pandemic and constrained supply-chain conditions), and I've had three different editors so I don't really know who did what and when, but my sincere thanks to Pete (Schiffer), my publisher, and to the entire publishing team—including Ann Charles, Tod Benedict, and Justin Watkinson—for all you have done!

While many individuals have assisted in ensuring the accuracy and comprehension of the manuscript, I alone am responsible for errors or omissions in the manuscript.

Last, but not least, thank you to YOU, dear reader. Without you, this book would have no audience.

APPENDIX

PIRKEI DE RABBI ELIEZER[265]

Presented below is the translated text (of the relevant sections), with only minor syntactical or grammatical modifications to enhance readability. Selected footnotes have been included, some modified or appended for brevity or clarity.

THE PRINCIPLE OF INTERCALATION[266]

On the 28th of Elul[267] the sun and the moon were created. The number of years, months, days, nights, hours, [268] terms, seasons, cycles, and intercalation were before the Holy One, blessed be He, and He intercalated the years and afterward He delivered the (calculations) to the first man in the garden of Eden, as it is said, "This is the calculation for the generations of Adam" (Genesis 5:1), the calculation of the world is therein for the generations of the children of Adam.

Adam handed on the tradition to Enoch,[269] who was initiated in the principle of intercalation, and he intercalated the year, as it is said, "And Enoch walked with G-d" (Genesis 5:22). Enoch walked in the ways of the calculation concerning the world, which G-d had delivered to Adam. And Enoch delivered the principle of intercalation to Noah,[270] and he was initiated in the principle of intercalation, and he intercalated the year, as it is said, "While the earth remaineth, seed-time and harvest, and cold and heat, and summer and winter" (Genesis 8:22). "Seed-time" refers to the Teḳuphah[271] of Tishri, "harvest" refers to the Teḳuphah of Nisan, "cold" refers to the Teḳuphah of Ṭebeth, and "heat" refers to the Teḳuphah of Tammuz; "summer" is in its season and "winter" is in its season.

The counting of the sun is by day and the counting of the moon is by night: "They shall not cease."

Noah handed on the tradition to Shem, and he was initiated in the principle of intercalation; he intercalated the years and he was called a priest, as it is said, "And Melchizedek King of Salem . . . was a priest of G-d Most High" (Genesis 14:18). Was Shem the son of Noah a priest? But because he was the firstborn, and because he ministered to his G-d by day and by night, therefore was he called a priest. Shem delivered the tradition to Abraham; he was initiated in the principle of intercalation and he intercalated the year, and he (also) was called priest, as it is said, "The Lord hath sworn, and will not repent, Thou art a priest for ever after the order of Melchizedek" (Psalms 110:4). Whence do we know that Shem delivered the tradition to Abraham? Because it is said, "After the order of Melchizedek" (ibid.). Abraham delivered the tradition to Isaac, and he was initiated in the principle of intercalation, and he intercalated the year after the death of our father Abraham, as it is said, "And it came to pass after

265. Pirkei de-Rabbi Eliezer is an aggadic-midrashic (nonlegalistic/philosophical) work on the Torah containing exegesis and retellings of biblical stories. There is unanimous scholarly agreement that the compilation of "Pirkei de-Rabbi Eliezer" is a work of the eighth or ninth century, most likely shortly after 830. Original authorship is unknown, but the presumed author is Rabbi Eliezer, son of Hyrkanos, ca. first to second century CE. Presented here is chapter 8 (or chapter 7 on the basis of the version of source manuscript), which examines the secrets of the calendar (Sod Ha-Ibbur). Translation by Gerald Friedlander, 1916, London.

266. Chapter 8 deals exclusively with the principle of intercalation and is the only chapter presented in this appendix. Chapter 7 of the same book discusses "The Course of the Moon," which readers may find interesting in relation to this topic.

267. I have tried to leave the translation entirely unaltered, excepting grammatical inconsistencies and some minor syntactic changes to assist twenty-first-century readers. As an example, modern reference to the twelfth month of the Jewish civil year (or sixth month of the ecclesiastical year)—Elul—is spelled with a singular l. British English of the early twentieth century was somewhat more archaic, with Elul spelled Elull.

268. Early editions of the manuscript included "hours."

269. The fact that, according to the Torah, Enoch lived 365 years (corresponding to the number of days in the solar year) is noteworthy.

270. This should probably read "Methuselah," and the text should continue "who handed it on to Noah."

271. Tekuphah means not only season, but also the time of solstice and equinox according to the season.

the death of Abraham, that G-d blessed Isaac his son" (Genesis 25:11), because he had been initiated in the principle of intercalation and had intercalated the year (therefore) He blessed him with the blessing of eternity. Isaac gave to Jacob all the blessings and delivered to him the principle of intercalation. When Jacob went out of the (Holy) Land, he attempted to intercalate the year outside the (Holy) Land. The Holy One, blessed be He, said to him: Jacob! Thou hast no authority to intercalate the year outside the land (of Israel); behold, Isaac thy father is in the (Holy) Land, he will intercalate the year, as it is said, "And G-d appeared unto Jacob again, when he came from Paddan-Aram, and blessed him" (Gen. 35:9). Why "again"? Because the first time He was revealed to him, He prevented him from intercalating the year outside the (Holy) Land, but when he came to the (Holy) Land, the Holy One, blessed be He, said to him: Jacob! Arise, intercalate the year, as it is said, "And G-d appeared unto Jacob again . . . and blessed him" (ibid.), because he was initiated in the principle of the intercalation, and He blessed him (with) the blessing of the world.

Jacob delivered to Joseph and his brethren the principle of intercalation, and they intercalated the year in the land of Egypt. (When) Joseph and his brethren died, the intercalations ceased from Israel in Egypt, as it is said, "And Joseph died, and all his brethren, and all that generation" (Exodus 1.6). Just as the intercalations were diminished from the Israelites in the land of Egypt, likewise in the future will the intercalations be diminished at the end of the fourth kingdom until Elijah, be he remembered for good, shall come. Just as the Holy One, blessed be He, was revealed to Moses and Aaron in Egypt, likewise in the future will He be revealed to them at the end of the fourth kingdom, as it is said, "And the Lord spake unto Moses and Aaron in the land of Egypt, saying, 'This month shall be unto you the beginning of months'" (ibid. 12:2). What is the significance of the word "saying"? Say to them, until now the principle of intercalation was with Me; henceforth it is your right to intercalate thereby the year.

Thus were the Israelites wont to intercalate the year in the (Holy) Land. When they were exiled to Babylon, they intercalated the year through those who were left in the (Holy) Land. When they were all exiled and there were not any (Jews) left[272] in the (Holy) Land, they intercalated the year in Babylon. (When) Ezra and all the community with him went (to Palestine), Ezekiel wished to intercalate the year in Babylon; (then) the Holy One, blessed be He, said to him: Ezekiel! Thou hast no authority to intercalate the year outside the Land; behold, Israel thy brethren, they will intercalate the year, as it is said, "Son of man, when the house of Israel dwell in their own land" (Ezekiel 36:17). Hence (the Sages) have said, "Even when the righteous and the wise are outside the Land, and the keeper of sheep and herds are in the Land, they do not intercalate the year except through the keeper of sheep and herds in the Land. Even when prophets are outside the Land and the ignorant are in the Land they do not intercalate the year except through the ignorant who are in the land (of Israel)," as it is said, "Son of man, when the house of Israel dwell in their own land" (ibid.) it is their duty to intercalate the year.

On account of three things is the year intercalated, on account of trees, grass, and the seasons (Teḳuphoth). If two of these (signs) be available and not the third, they do not intercalate the year, (that is to say) neither because of the trees nor because of the grass. If one (sign) be available and the other two be absent, they do not intercalate the year on account of the Teḳuphoth.[273] If the Teḳuphah of Ṭebeth had occurred on the 20th day of the month or later, they intercalate the year; but until the 20th day of the month Ṭebeth or earlier, they do not intercalate the year.

The cycle of intercalation is 19 years, and there are 7 small cycles therein; some of these are (separated by) 3 years, some (by) 2 years, others (are separated by) 3 or 2 years, or (by) 3, 3, and 3 years, (the order of the cycles being) 3rd, 6th, 8th, 11th, 14th, 17th, and 19th years. There are two (sets) of three years' cycles.[274]

272. After the murder of Gedaliah; see T.B. Sabbath, 145B.

273. "This is an error," says Luria. "It should be the 16th"; see Babylonian Talmud Rosh Hashanah, loc.cit., and Babylonian Talmud Sanhedrin, 13A: for if the Tekuphah of Tebeth fell on the 21st of Tebeth, then the Tekuphah of Nisan would be on the 24th of Nisan (ninety-one days' interval), which is the day after Passover; accordingly Passover would not be in Abib (the Tekuphah in Nisan) and therefore Adar Sheni should be intercalated. The reading in our text (the 20th) is approved by Schwarz (Der judische Kalender, page 36, note 3). The "Megillah of Abiathar" (op. cit. page 471) reads: "If the Tekuphah of Tebeth had occurred from half (of the month) and later they intercalate the year, but until half (of the month) and earlier they do not intercalate the year." The printed editions read: "If the Tekuphah had occurred by the 20th day of the month or earlier they intercalate the year; but from the 20th day of the month or later they do not intercalate the year." This is clearly wrong. The correct reading is preserved by our manuscript, which is confirmed by the Oxford manuscript. (d. 35). The rule in our text does not apply now in actual practice. See Jozeroth, ed. Arnheim, page 73.

274. According to the Oxford manuscript (d. 35), the text should read thus: the 3rd, 5th, 8th, 11th, 14th, 16th, and 19th, agreeing with the cycle of Meton the Greek astronomer, with the exception that the latter has the 13th year instead of the 14th year.

The intercalation takes place in the presence of three; Rabbi Eliezer says that ten (men are required), as it is said, "G-d standeth in the congregation of G-d" (Ps. 82:1), and if they become less than ten, since they are diminished they place a scroll of the Torah before them, and they are seated in a circle in the courtroom, and the greatest (among them) sits first, and the least sits last; and they direct their gaze downward to the earth and (then) they stand and spread out their hands before their Father who is in heaven, and the chief of the assembly proclaims the name (of G-d), and they hear a Bath Ḳol (saying) the following words: "And the Lord spake unto Moses and Aaron . . . saying, 'This month shall be unto you'" (Ex. 12:1, 2).

If, owing to the iniquity of the generation, they do not hear anything at all,[275] then, if one may say so, He is unable to let His glory abide among them. Happy were they who stood in that place in that hour, as it is said, "Happy is the people who know the joyful sound: they walk, O Lord, in the light of thy countenance" (Psalm 89:15); in the light of the countenance of the Holy One, blessed be He, they walk.

On the New Moon of Nisan the Holy One, blessed be He, was revealed to Moses and Aaron in the land of Egypt, and it was the 15th year of the great cycle of the Moon, the 16th year of the cycle of intercalation, (and He said): "henceforward the counting devolves on you."

275. Rabbi Eliezer was permitted to hear the Bath Kol. See Babylonian Talmud Baba Mezia, 59B, and Babylonian Talmud Sotah, 48B. In later times, this privilege was withdrawn because of the sins of the people.

INDEX

C

D

E

H

I

J

N

O

P

Q

R

S